Twenty Ninth International Conference on Surface Modification Technologies 2015

Kongens Lyngby, Denmark
10-12 June 2015

Editors:

T.S. Sudarshan **M.A.J. Somers**

ISBN: 978-1-5108-2791-2

Printed from e-media with permission by:

Curran Associates, Inc.
57 Morehouse Lane
Red Hook, NY 12571

Some format issues inherent in the e-media version may also appear in this print version.

Printed by Curran Associates, Inc. (2016)

Valardocs
43 Sarangapani St.
Chennai 600017
Tamil Nadu, India

Phone: 91 44 28342521

valardocs@gmail.com

Additional copies of this publication are available from:

Curran Associates, Inc.
57 Morehouse Lane
Red Hook, NY 12571 USA
Phone: 845-758-0400
Fax: 845-758-2633
Email: curran@proceedings.com
Web: www.proceedings.com

SURFACE MODIFICATION TECHNOLOGIES XXIX

EDITED BY T.S. SUDARSHAN AND M.A.J. SOMERS

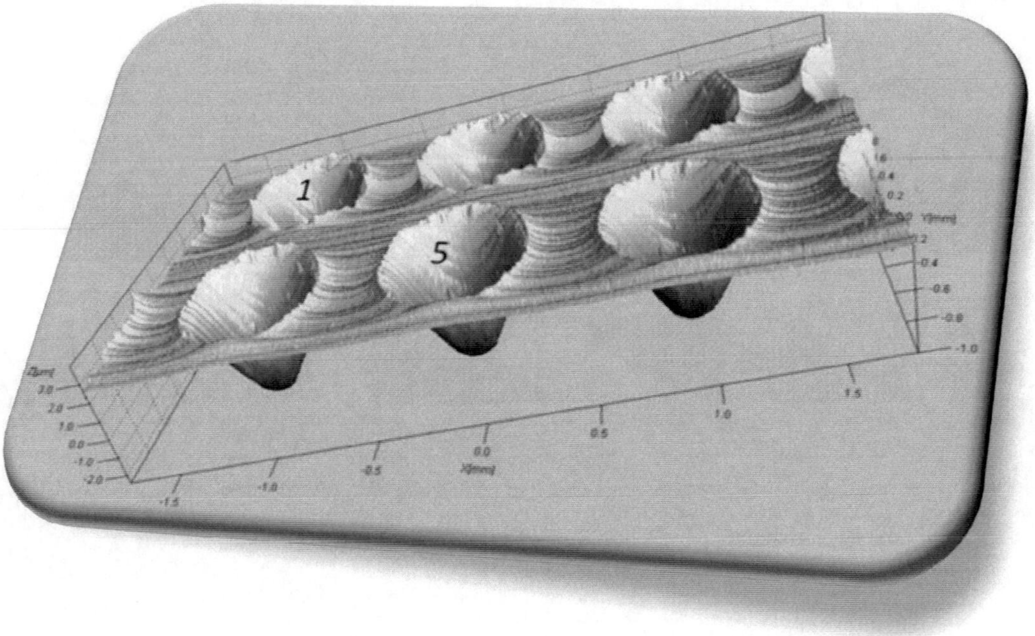

Surface Modification Technologies XXIX

Proceedings of the Twenty Ninth International Conference on Surface Modification Technologies

Held at

Technical University of Denmark

Kongens Lyngby, Denmark

June 10-12, 2015

Editors (Proceedings)

Dr. T. S. Sudarshan
President and CEO
Materials Modification, Inc.
2809-K Merrilee Drive
Fairfax, VA - 22031
U.S.A.

Prof. Marcel A.J. Somers
Section Head
Department of Mechanical Engineering
Materials and Surface Engineering
Technical University of Denmark
Produktionstorvet
Building 425, room 120
2800 Kgs. Lyngby
Denmark

VALARDOCS

Contents

Surface Modification Technologies XXIX

Proceedings of the Twenty Ninth International Conference on Surface Modification Technologies

Held at

Technical University of Denmark

Kongens Lyngby, Denmark

June 10-12, 2015

Editors (Proceedings)

Dr. T. S. Sudarshan
President and CEO
Materials Modification, Inc.
2809-K Merrilee Drive
Fairfax, VA - 22031
U.S.A.

Prof. Marcel A.J. Somers
Section Head
Department of Mechanical Engineering
Materials and Surface Engineering
Technical University of Denmark
Produktionstorvet
Building 425, room 120
2800 Kgs. Lyngby
Denmark

Foreword

The Twenty Ninth (29th) International Conference on ***Surface Modification Technologies*** (SMT28) was held in the Technical University of Denmark, Kongens Lyngby, Denmark, between June 10th – 12th, 2015. The conference had the primary objective of bringing together an interdisciplinary community of applied engineers, industrialists, scientists and academicians to foster discussions on progress in the areas of surfaces, coatings, and their preparation, modification and properties. The meeting was small, friendly, and collaborative with many mutual exchanges and numerous relationships were established between diverse groups. Smaller meetings tend to have a much higher level of personal interaction leading to significant advancements through exchanges, joint proposals and long term friendships and even joint publications. Over the years, numerous exchanges have taken place in both research and visits and co-publication of papers that have brought together different viewpoints and thereby strengthened the knowledge of many through the significant contributions to the archival literature.

This book is a compilation of most of the papers that were presented. The papers in the proceedings reflect significant work by the authors and were peer reviewed and corrected prior to publication. We would like to thank the numerous reviewers who gave generously of their time and also reviewed on priority at times and for their dedication to help this proceedings come to print. Thanks are also due to the several authors for their willingness to adapt to the requirements of this publication.

We would also like to acknowledge the generous use of the communication facilities and time realized by our employers in the compilation of this volume that ensured its publication.

Dr. T. S. Sudarshan
President and CEO
Materials Modification, Inc.
2809-K Merrilee Drive
Fairfax, VA - 22031
U.S.A.

Prof. Marcel A.J. Somers
Section Head
Department of Mechanical Engineering
Materials and Surface Engineering
Technical University of Denmark
Produktionstorvet
Building 425, room 120
2800 Kgs. Lyngby
Denmark

Electropolishing of Ti6Al4V Alloy Fabricated by Selective Laser Melting

M.B. García-Blanco*, O. Garrido, G. Vara, J.A. Díez
Fundación Cidetec. Parque Tecnológico de San Sebastián
P° Miramón, 196 20009 Donostia-San Sebastián (Gipuzkoa), Spain.
*Email: bgarcia@cidetec.es

Abstract

Selective Laser Melting (SLM) is an additive manufacturing technology that creates products by adding layers of material, one on top of the other, until the physical object is created. SLM is an extraordinary tool as it offers the possibility to build strong, flexible and rigid metallic structures that are still light in weight. Usually parts produced by this technology require a secondary processing to achieve the final surface finishing and properties. For this purpose several post-process methods such as mechanical polishing, abrasive flow polishing or chemical milling have been used. However, for most of these surface treatments, geometrical complexity of the treated parts is the major challenge. In this context, electropolishing (EP) is recommended since it allows the polishing of difficult geometries, leading to a bright, clean and smooth appearance in a short time. In this work, sand-blasting, EP and the combination of both processes were used to improve the surface roughness of SLM generated Ti6Al4V parts. The obtained results showed that the combination of sand-blasting with EP leads to the smoothest and brightest surface, free from contaminants, even in difficult-to-access areas of the analysed 3D component.

1.0. Introduction

The reduction of energy consumption in the aerospace industry leads to social, environmental and economic benefits and will be important to ensure its sustainability. Fuel economy in aircraft can be achieved, among others, through the reduction of the weight of an aircraft. For this purpose, it is important to achieve the topological optimization of parts and structures in aircrafts, which must be manufactured in lightweight materials such as Ti6Al4V, reaching large reductions in weight and structural mass, but maintaining their properties and mechanical requirements. In this context, Additive Manufacturing (AM) technologies such as Selective Laser Melting (SLM), through the improvement of the components design, can provide novel structures with these features. SLM uses high powered laser to melt metallic powders together, layer upon layer, to shape the product. Due to the high degree of freedom given by this technology, parts with extremely complex shape and geometry can be produced. Nevertheless, one of the main drawbacks of this technology is the resulting high surface roughness which is mainly determined by the powder size distribution and also by other processing parameters like the thickness of the processed powder layer or the scanning speed [1,2].

Several post-process methods such as mechanical polishing, abrasive flow polishing or chemical milling have been used to enhance the surface quality of the final parts produced by SLM[3–5]. For most of these surface treatments, geometrical complexity of the treated parts is the major challenge. For instance, mechanical polishing such as sand-blasting is difficult to apply on internal structures. Blasting process consists on the impact of a jet of abrasive particles against the surface by compressed air; this could increase the surface reactivity of the metal and the blasting particles can induce abrasive pollution on the surface. Abrasive flow polishing could only achieve the desired surface finish on particular areas due to the directional flow of the abrasive fluid and chemical milling usually does not completely eliminate surface features due to the isotropic etching effect. In this context, electropolishing (EP) is recommended as a post-process method for components produced by SLM since it allows the polishing of difficult geometries leading to a bright, clean and smooth appearance in relatively short time.

EP is a surface finishing process in which the surface roughness is eliminated through a selective electrochemical dissolution. The metal to be polished is immersed in an electrolyte, which must be carefully chosen for each system, anodically polarized and dissolved by applying a specific amount of current for a fixed period of time. When dealing with titanium the process normally uses toxic, acidic and aggressive electrolytes. The reason for this choice is that titanium readily forms an oxide layer in the presence of oxygen. Therefore, very strong acids such as perchloric, hydrofluoric, and/or methanolic with sulphuric acids must be used as electrolytes to reduce the formation of the oxide layer[6–8]. In order to improve the safety of the process, nowadays, different strategies are being considered. On the one hand, the application of bipolar current pulses during the process[9] and, on the other, the use of less dangerous non-conventional electrolytes like ionic liquids or alcoholic solutions[10,11].

In this context, this work has been focused on the application of an EP process to reduce the surface roughness of a Ti6Al4V component fabricated by SLM and to compare the results with that obtained by a traditional mechanical polishing (MP) process (sand-blasting), and with the combination of both techniques (MP + EP).

In order to avoid the use of hazardous electrolytes, in this study the electropolishing process has been carried out in an alcoholic solution based on ethyl alcohol and isopropyl alcohol.

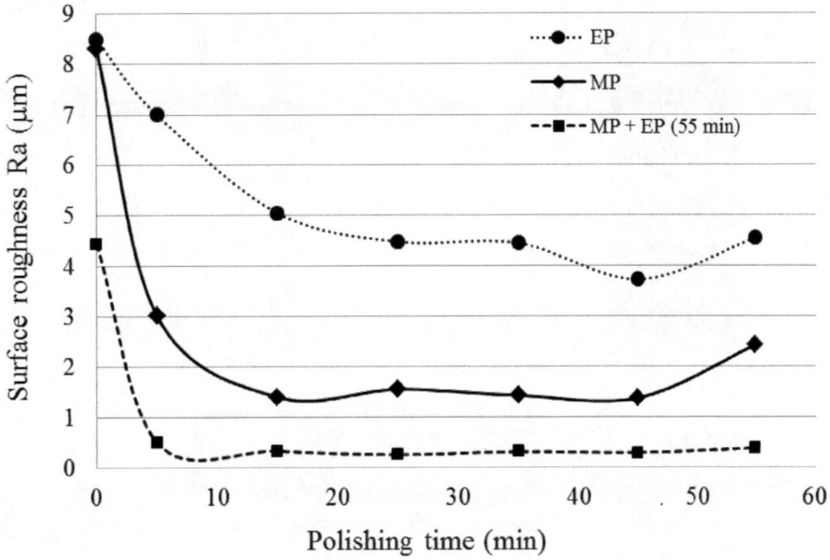

Fig. 1: Average surface roughness (Ra) after electropolishing (EP), sand-blasting (MP) and sand-blasting + 55 min of electropolishing

The investigation has been carried out first in 2D components, to compare the effect of EP, MP and EP+MP on the surface roughness and to define the parameters that lead to the smoothest surface in each of the analysed technologies. Then, processes have been scaled up to 3D components in order to evaluate the effect of EP, MP and EP+MP in the polishing of a more complex structure.

2.0. Experimental Details

The material used for the experiments was a Ti6Al4V alloy fabricated by SLM and provided by CATEC AERO in two different geometries (a planar 2D piece and a tri-dimensional component).

EP experiments were performed applying a voltage of 38 V with an EA-PS 5040-40A power-supply, using a two electrode system in a 400 mL electrolyte bath at 30 °C. The anode was the Ti6Al4V piece, the cathode was a CP titanium foil and the distance between them was set in 3 cm. The EP process time was varied from 5 to 55 min. A non-aqueous electrolyte comprising ethyl alcohol (700 ml/L), isopropyl alcohol (300 ml/L), aluminum chloride (60 g/L), and zinc chloride (250 g/L) was used[12]. Magnetic stirring was employed at the bottom of the electrolytic cell in order to maintain a control hydrodynamic of the electrolyte. Subsequent to EP, substrates were rinsed with distilled water and dried with air.

Ti6Al4V samples were sandblasted using round shape glass microspheres of 40-70 µm in a PALLINATRICI NORBLAST Bologna-Italy (S9 Model) sand blaster, using a nozzle diameter of 8 mm. The distance between sample holder and nozzle was set in 50 mm and the pressure drop across the nozzle was established in 5.5 bar. The process length was varied from 5 to 55 min.

The study of the effect of the combination of both processes was carried out applying sand-blasting with different process lengths (from 5 to 55 min) followed by 55 min of EP.

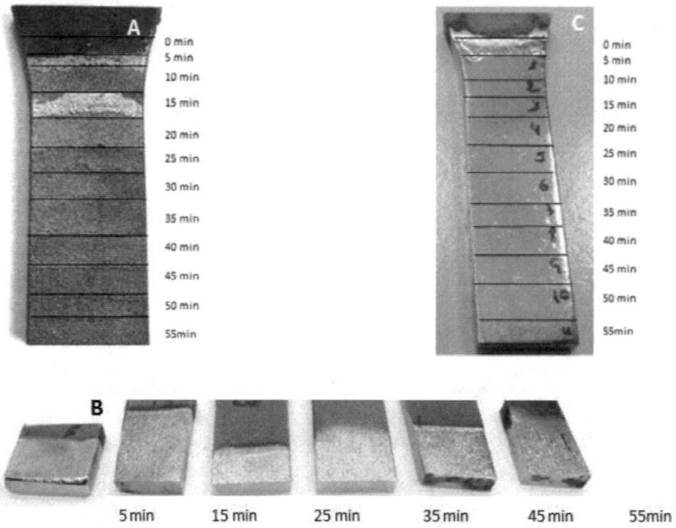

Fig. 2: Ti6Al4V samples after **(a)** mechanical polishing **(b)** electropolishing and **(c)** mechanical polishing + 55 min of electropolishing

A TAYLOR-HOBSON Talysurf Intra 50 mm profilometer was used to measure the surface roughness of the substrates. The surface roughness was calculated by the averaged Ra values (average of the distances between the roughness profile and the intermediate line of the measurement length) and also by measuring the % of roughness decrease (RD) which was determined by the Eq. 1

$$\% \, RD = \frac{Ra_{t=0} - Ra_{t=t}}{Ra_{t=0}} * 100 \quad \ldots\ldots\ldots\ldots\ldots\ldots(1)$$

The morphological characterization of the surfaces was carried out with a JSM-5500LV model JEOL Scanning Electronic Microscope. This microscope is coupled with a 200 Series INCA ENERGY lead from OXFORD INSTRUMENTS, for the qualitative analysis of the treated surface composition.

3.0. Results and Discussion

3.1. Surface treatments on SLM generated 2D Ti6Al4V samples

Fig. 1, shows the arithmetic average surface roughness, Ra, for untreated, sandblasted and electropolished samples, and also for the samples where both processes were combined.

When EP is applied, surface roughness decreases with polishing time. After 25 min the process stabilizes and there are no significant improvements in surface roughness. This tendency is characteristic of the EP process and is been described by other authors in other materials such as stainless steel[13].

A similar trend can be observed when sand-blasting is applied. Varying the sand-blasting process length, the surface finish is improved over time and eventually reached a stationary level at 15 min. This tendency is also representative of the sand-blasting process and has been described elsewhere[14,15].

The analysis of the results in Fig. 1 indicates that sand-blasting provides a much smoother surface than EP. While, for example, 25 min of EP reduce the surface roughness to around 50%, the same sand-blasting time leads to a surface roughness decrease of 80%.

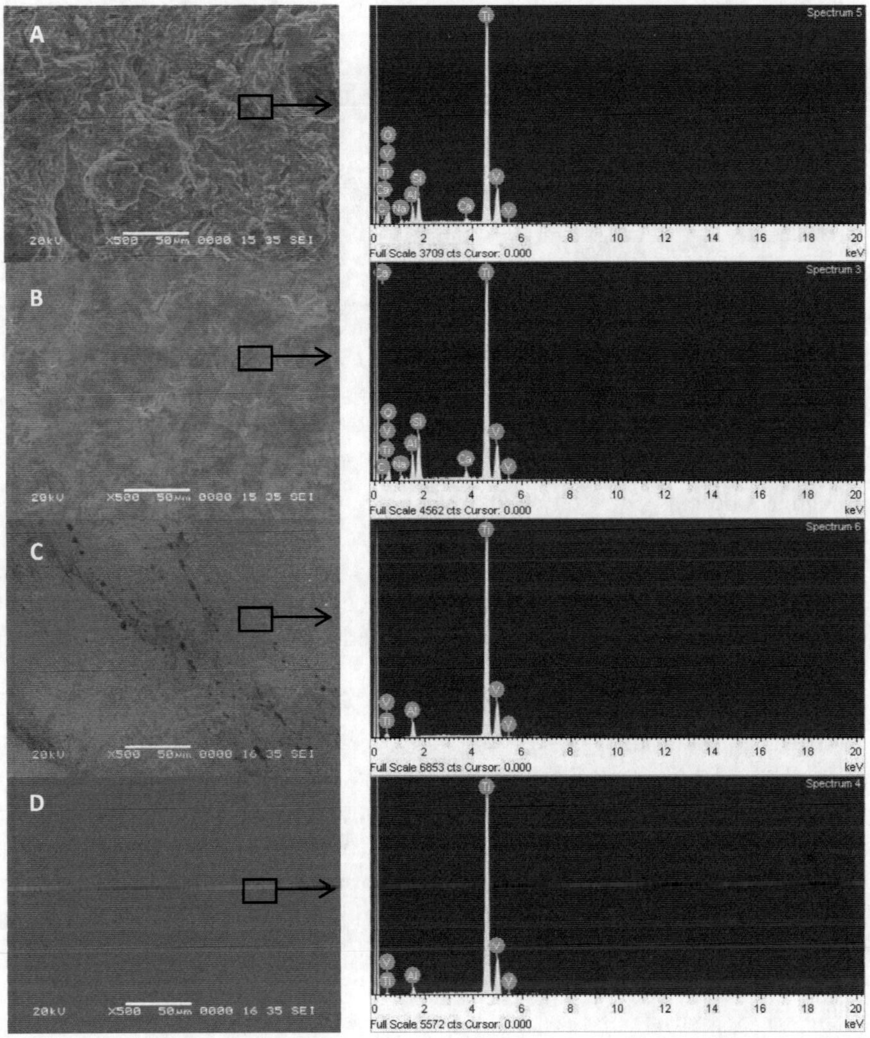

Fig. 3: SEM micrographs and EDX spectra of **(a)** as-received **(b)** mechanically polished **(c)** electropolished **(d)** mechanical polished + electropolished samples

The relatively high value of surface roughness for the EP pieces is due to the partially melted particles fused with the surface, that are not fully removed by the electrochemical oxidation process. If the surface is pre-treated with sand-blasting before EP, regardless the sand-blasting process length, these particles can be removed or flattened, hence making the electrochemical process much more efficient. This behavior was also observed by L. Lober et al. in stainless steel[3].

The images of the resulting surfaces are illustrated in Fig. 2. As can be seen, after sand-blasting (Fig. 2(a)) the pieces look dull with little change on the aspect from the as-received surface. In contrast, electropolished surfaces (Fig. 2(b)) are bright with metallic appearance, achieving the greater degree of gloss at 55 min. High quality mirror-like finish can be observed in Fig. 2(c) where the combination of both surface treatments has been applied.

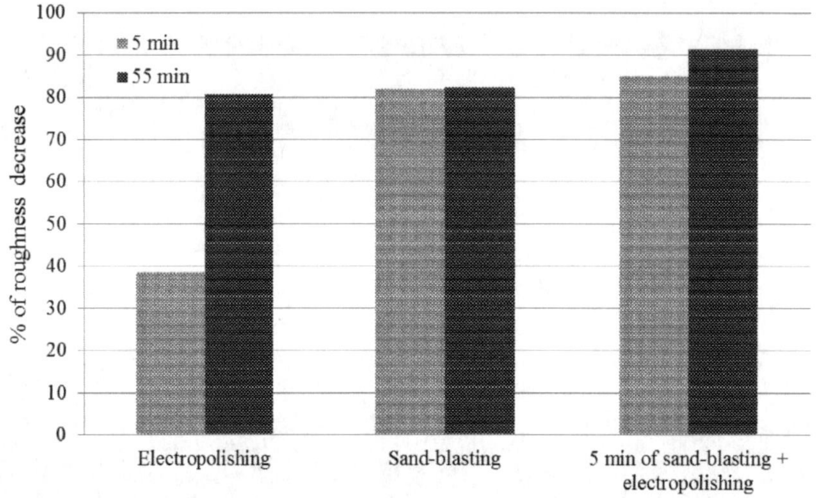

Fig. 4: % of roughness decrease after different surface treatments applied to the as-received Ti6AL4V pieces

SEM images and EDX spectra of some of the treated surfaces are shown in Fig. 3. It could be observed from the micrographs that the as-received samples (Fig. 3(a)) have an uneven and concave surface. The analysis of the chemical composition of this surface indicates that there are some impurities like Na, Si and Ca, related to the mechanical polishing process applied just afterwards SLM. On the other hand, SEM micrograph after one of the sand-blasting experiments (Fig. 3(b)) shows a less textured surface with smaller but remaining few cavities. The EDX analysis of the blasted surface reveals that there are also chemical impurities due to the glass microspheres used during the process. By contrast, after EP (Fig 3(c)) the sample´s surface appears to be flat even though some defects are appreciated. Here, as shown in the EDX spectra, the surface is free from contaminants as only the elements of the alloy are detected. These results are also observed in the surfaces achieved after the combination of sand-blasting and EP (Fig. 3(d)), however, in this case, the surface is smooth and uniform and free from defects.

The analysis of the obtained results for the 2D pieces reveals that the application of sand-blasting leads to a great decrease in surface roughness in very short process times. Nonetheless, the appearance of the treated pieces is dull and the surface is contaminated by the glass microspheres used during the process.

From the above it is clear that the combination of sand-blasting with EP leads to very smooth and bright surfaces, free from contaminants, in flat pieces. Nevertheless, taking into account that the SLM technology is commonly used to create complex geometries, in this work, also the evaluation of the effect of these processes, alone and in combination, in non-planar surfaces has been carried out.

3.2. Surface treatments on SLM generated 3D Ti6Al4V samples

In order to evaluate how sand-blasting, EP and the combination of both processes affect the surface of a non-planar geometry, different experiments have been carried out using the previously studied shortest and longest process times of each technology. Even though shorter electropolishing times had led to the same surface roughness as the longest one, the last has been chosen due to the higher degree of gloss of the electropolished surfaces.

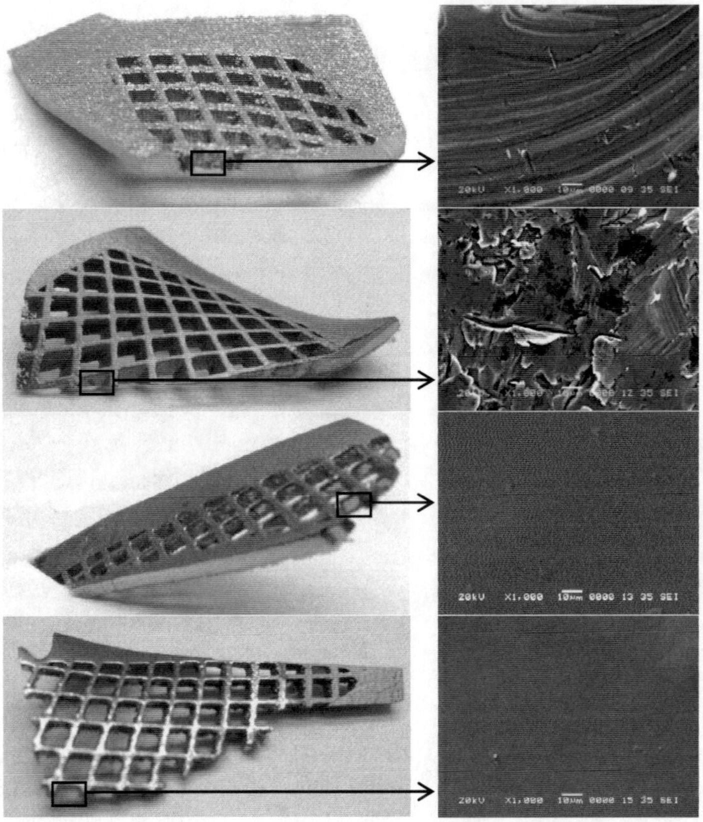

Fig. 5: Images and SEM micrographs of **(a)** as-received **(b)** sand-blasted **(c)** electropolished **(d)** sand-blasted + electropolished 3D pieces

Fig. 4 gives the % of roughness decrease reached on the experiments. As can be observed, for the EP experiments the longest process time has led to the best surface finish. In contrast, for the sand-blasting experiments, similar results have been obtained for both process lengths. As previously observed, the combination of 5 min of sand-blasting followed by EP has led to the highest surface roughness decrease, achieving more than a 90% with the longest EP time.

With the aim of investigating how the difficult-to-access surfaces of the evaluated pieces have been polished, SEM images of these areas on as-received, sand-blasted, electropolished and sand-blasted followed by electropolished surfaces have been analyzed (Fig. 5). The comparison between the morphology of the non-planar surface of the as-received piece (Fig. 5(a)) and the same area of a sand-blasted piece (Fig. 5(b)) reveals that the last surface has been affected by the process. Nevertheless, the uneven surface obtained after sand-blasting indicates that this process is not able to polish effectively non-planar areas of a 3D geometry.

On the other hand, the non-planar surfaces of the pieces where the EP process has been applied, either alone (Fig. 5(c)) or combined with sand-blasting (Fig. 5(d)), are smooth and homogeneous. These results evidence that EP can be successfully used to polish more difficult-to-access surfaces.

4.0. Conclusions

In this work, sand-blasting and EP processes were used to improve the surface roughness of SLM generated Ti6Al4V parts. For this purpose, different polishing times were analyzed applying both processes, alone and in combination, in 2D and 3D geometries. The comparison of the different surface treatments applied led to the next conclusions:

- Sand-blasting is a good polishing method to reduce surface roughness of Ti6Al4V pieces fabricated by SLM but it has the disadvantage of introducing contaminants on the surface. Moreover, this process leads to a dull finish and is not able to polish effectively all the areas of a given piece.
- EP can reduce the surface roughness of a SLM generated Ti6AlV component up to an 80 %, even in difficult-to-access areas of the analyzed 3D piece. The resulting surface is smooth, bright and free from impurities.
- The combination of sand-blasting with EP leads to the smoothest and brightest surface, free from contaminants, even in difficult-to-access areas of the analyzed 3D piece.

5.0. Acknowledgments

The authors gratefully acknowledge CATEC (Center for Advances Aerospace Technologies, Seville, Spain) for providing the SLM generated Ti6Al4V pieces.

6.0 References

1. Yadroitsev, I.; Smurov, I.; Physics Procedia, 2011, vol. 12(1), p. 264.
2. Yadroitsev, I.; Bertrand, P.; Smurov, I.; Appl. Surf. Sci., 2007, vol. 253(19), p. 8064.
3. Löber, L.; Flache C.; Petters, R.; Kühn, U.; Eckert,; Rapid Prototyping J., 2013, vol. 19(3), p. 173.
4. Rossi S.; Defloriann, F.; Venturini, F.; J. Mater. Process. Technol. 2004, vol. 148, p. 301.
5. Samoya, M.; Cormier, D.; Harrysson, O.; Cansizoglu, O.; Mechanical Properties of Chemically Milled Titanium Lattice Structures, 2009.
6. Chen, C.C.; Chen, J. H.; Chao, C.G.; J. Mater. Sci., 2005, vol. 40, p. 4053.
7. Neelakantan, L.; Hassel, A.W.; Electrochimica Acta, 2007, vol. 53, p. 915.
8. Khun, A.; Metal Finishing, 2004, vol. 103(6), p. 80.
9. Inman, M.; Hall, T.; Garich, H.; Taylor, E.J.; Environmentally benign electropolishing for biomedical alloys, ASM 2014 Int. Manufact. Sci. and Eng. Conf. collocated with the JSME 2014 Int. Conf. on Mat. and Procc. And the 42nd North American Manufact. Research Conf. Processing, 2014, vol. 2, p. 9.
10. Abbott, A.; Ryder, K.S.; Konig, U.; Transactions of the Institute of Metal Finishing, 2008, vol. 86(4), p. 196.
11. Uda, T.; Tsukimoto, K.; Nakagawa, H.; Murase, K.; Nose, Y.; Awakura, Y.; Mater. Transac., 2011, vol. 52(11), p. 2061.
12. Tajima, K.; Hironaka, M.; Chen, K. -K.; Nagamasu, Y.; Kakigawa, H.; Kozono, Y.; Dental Mater. J., 2008, vol. 27(2), p. 258.
13. Núñez, P.J.; García-Plaza, E.; Hernando, M.; Trujillo, R.; Characterization of surface finish of electropolished stainless steel AISI 316L with varying electrolyte concentrations, Proceedings of the 5th Manufact. Eng. Soc. International Conference – Zaragoza, 2013
14. Qu, J.; Shih, A.J.; Scattergood, R.O.; Luo, J.; J. Mater. Process. Technol., 2005, vol. 166, p. 440.
15. Monetta, T.; Belluci, F.; Open J. of Rgenerative Medicine, 2012, vol. 1(3), p.41.

Surface Engineering to Mitigate High Temperature Corrosion

U. Kamachi Mudali*, Ch. Jagadeeswara Rao, A. Ravi Shankar, K. Thyagarajan, E. Vetrivendan, C. Mallika
Corrosion Science and Technology Group, Indira Gandhi Centre for Atomic Research, Kalpakkam – 603 102, India.
*Email: kamachi@igcar.gov.in

Abstract

Pyrochemical reprocessing in molten salt (LiCl-KCl) medium is proposed for the reprocessing of spent metallic fuels discharged from future sodium cooled Fast Breeder Reactors (FBRs) in India. Structural materials for pyrochemical reprocessing are expected to exhibit high corrosion resistance in molten salt and molten metal and/or alloy (U, U-Zr, U-Zr-Pu) environment in the temperature range 500-1400°C. For this purpose, the corrosion behaviour of stainless steels, Cr-Mo steels, Ni based alloys and high density graphite was investigated in molten salt and molten uranium. To mitigate corrosion, yttria stabilized zirconia (YSZ), alumina-40% titania and yttria coatings were deposited on these materials with and without bond coats by atmospheric plasma spray process and the coated samples were corrosion tested in molten salt and molten uranium. The coated samples exhibited superior corrosion resistance in molten salt and uranium, compared to the uncoated samples. The durability of the coatings was established by performing thermal cycling studies.

1.0. Introduction

For the reprocessing of spent U-Pu-Zr based metallic fuels discharged from future Fast Breeder Reactors (FBRs) in India, pyrochemical reprocessing in molten salt (LiCl-KCl) medium is proposed[1]. All the unit operations of pyrochemical reprocessing employ either molten LiCl-KCl salt or molten metal (uranium) and/or the alloy U-Pu-Zr at temperatures ranging between 500 and 1400°C[2]. Three of the major operations in pyrochemical reprocessing are (i) Electrorefining of the fuel elements and other actinides using molten LiCl-KCl eutectic salt as the electrolyte in an electrorefiner at 500°C,[3] (ii) Distillation of cadmium and entrained salt from the electrodeposit and consolidation of the actinides (U, Pu) in a cathode processor[4] at about 1400°C followed by (iii) Casting of the consolidated alloy form to fabricate fresh fuel pins. Structural materials considered for pyrochemical reprocessing are expected to have high corrosion resistance in molten salt and molten uranium. Various structural materials have been proposed for different unit operations of pyrochemical reprocessing[5]. Type 316L stainless steel (SS), nickel-based alloy (Inconel 600) and high density graphite (HDG) are proposed for molten LiCl-KCl salt purification process, while Cr-Mo steels and SS 410 are considered for electrorefiner vessel, anode basket, solid cathode, and liquid cadmium cathode crucible in the electrorefining process. High density graphite is chosen as the structural material for the fabrication of crucibles for salt preparation, consolidating the cathode deposit and for casting the deposit into fuel pins. Though graphite is a high melting material exhibiting excellent corrosion and thermal shock resistance, it undergoes severe oxidation in atmospheric air at temperatures above 500°C. As temperatures above 1400°C are envisaged for uranium melting, it is essential to protect the graphite crucible from oxidation and prevent undesirable reactions between reactive liquid uranium and graphite, by providing a protective and non-reacting ceramic coating. Similarly, metallic structural materials proposed for various components will also react with molten salt at high temperatures and undergo significant corrosion and degradation. Hence, in the present study, the compatibility of various structural materials and graphite (with and without proper ceramic coatings) with LiCl-KCl molten salt and molten uranium was evaluated at temperatures in the range 500-1400°C. Yttria coating was deposited on HDG substrate (with and without a proper bond coat) by plasma spraying and its corrosion behaviour in molten uranium and the durability of the coating during thermal cycling were also investigated for cathode processing applications.

2.0. Methods

2.1. Immersion studies

The LiCl–KCl eutectic salt was purified by passing chlorine gas through the mixture at 400°C for 15 min. The cell setup with this purified salt and the test samples were loaded in the molten salt test assembly (MOSTA). The 316L SS samples were tested at 600°C for 25, 100 and 250 h, and yttria stabilized zirconia (YSZ) coated 316L SS coupons were tested for 1000 h under purified argon gas environment. Molten salt immersion studies were carried out on (a) 9Cr-1Mo steel for 25 and 100 h durations, (b) 2.25Cr-1Mo steel for 25, 100 and 250 h and (c) Ni based alloys for 25, 50, 100, 250 and 500 h durations. Yttria stabilized zirconia coated 9Cr-1Mo steel and SS 304L samples and different forms of graphite were also tested up to 2,000 h. After the corrosion test, samples were removed and cleaned with water and acetone.

Table 1: Typical plasma spray parameters used for the development of coatings on structural materials

Process parameters (Units)	Cr_3C_2-NiCr	NiCrAlY	A40T	YSZ	Yttria
Plasma arc current (A)	470	500	470	600	600
Arc voltage (V)	50	75	50	64-70	70
Primary gas	Ar	Ar	Ar	Ar	Ar
Secondary gas	H_2	H_2	H_2	H_2	H_2
Powder feed rate (g/min)	50	60	50	46	45
Spray distance (mm)	100	150	100	100	100
Argon gas flow rate (l/min)	80	96	80	38	97
Carrier gas flow rate (l/min)	37	37	40	37	37

2.2. Development of coating

Coatings were developed with various ceramics by atmospheric plasma spray (APS) process on different structural materials for evaluating their corrosion behaviour in molten LiCl-KCl salt and molten uranium. The ceramic coatings YSZ, alumina-40% titania (A40T) and yttria along with the bond coats Ni-Cr, Cr_3C_2-25NiCr and NiCrAlY were produced using METCO 9MB, 3MB and Metallisation SG-100 plasma torches. Prior to developing top ceramic coating, an intermediate suitable bond coat was deposited on the substrate in order to increase the bond strength and adhesion of the ceramic coating on the substrate. The plasma spray parameters used for developing different ceramic coatings are given in Table 1.

2.3. Uranium melting studies

Uranium melting experiments were carried out in an induction melting setup housed in argon atmosphere glove box. Uranium was taken in the corrosion-tested miniature crucible and heated to 1350°C at the rate of 90°C/min. and soaked at this temperature for about 20 min. to ensure complete melting of uranium, and subsequently cooled down to room temperature.

2.4. Thermal cycling studies

Thermal cycling studies were performed using disc shaped specimens coated on only one face and the coated samples were vacuum sealed in quartz ampoules to prevent the oxidation of the uncoated surfaces. In the thermal cycling furnace, the samples were soaked for 10 min. at the required temperature followed by forced air cooling for 10 min. to ambient temperature. For yttria coated HDG samples with and without bond coat, thermal cycling experiments were carried out under argon atmosphere with the dwelling time of 1 h, followed by furnace cooling in a high temperature tubular furnace.

2.5. Characterization

Initial weight and dimensions of the samples were recorded before and after the corrosion test. Surface morphology and cross section of the coatings were characterized by scanning electron microscopy (SEM) coupled with Energy dispersive X-ray spectroscopy (EDX). Phase identification of the coatings before and after the corrosion test was carried out by X-ray diffraction (XRD) technique and Laser Raman spectrometry.

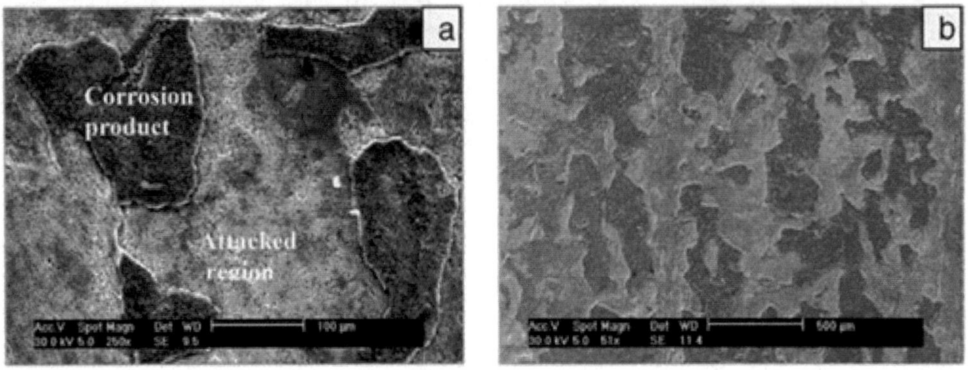

Fig. 1: SEM images of 316L SS exposed to molten LiCl-KCl salt for **(a)** 25 h **(b)** 100 h

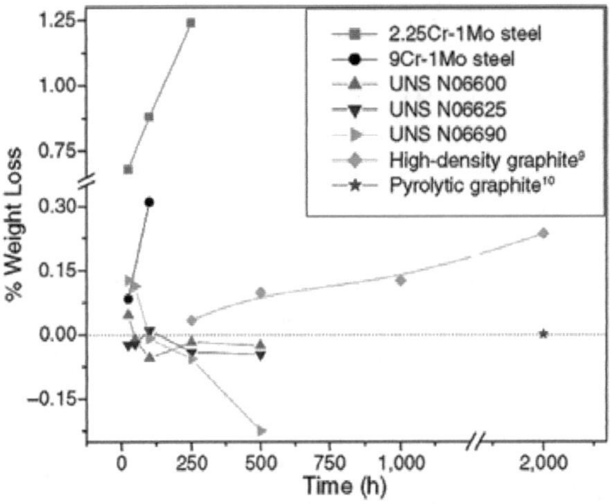

Fig. 2: Percentage weight loss of candidate materials exposed to molten salt at 600°C

3.0 Results and Discussion

3.1. Corrosion behaviour of structural materials in LiCl-KCl molten salt

The percentage weight loss of the uncoated 316L SS samples exposed to molten salt at 600°C for 25, 100 and 250 h durations were 0.3, 0.45 and 2.83, respectively, while the percentage weight loss of YSZ coated 316L SS exposed for 1000 h was only 0.08, indicating that the corrosion attack on YSZ coated 316L SS was insignificant[6]. After immersion for 25 and 100 h, the surface morphologies (Figs. 1(a) and 1(b)) of the uncoated 316L SS samples showed two distinct regions, the corrosion product and attacked surface. The attacked region indicates the porous nature of the surface. The EDX spectra of the uncoated 316L SS immersed in the molten salt for 100 h revealed the corrosion product region to be rich in Cr, while that of attacked region showed the presence of Fe, Ni and depletion of Cr[6]. It is apparent from the percentage weight loss (2.83 after 250 h) that the depletion of Cr was very rapid with increase in the duration of exposure. The percentage weight loss of materials exposed to molten LiCl-KCl salt at 600°C for various durations are shown in Fig. 2. It is evident that the corrosion attack on 2.25Cr-1Mo steel was more than that of 9Cr-1Mo steel[7].

Fig. 3: SEM micrographs from corrosion product region of **(a)** 2.25Cr-1Mo steel and **(b)** 9Cr-1Mo steel, exposed to molten LiCl-KCl salt for 100 h duration

The Ni-based alloys Inconel 600, Inconel 625, and Inconel 690 exhibited a decreasing trend of weight loss with exposure time.

The surface morphologies of molten salt corrosion tested 2.25Cr-1Mo and 9Cr-1Mo steels (Figs 3(a) and 3(b)) exhibited faceted grains and a porous matrix adjacent to the corrosion product region. Preferential leaching of Cr from the surface and the formation of Cr-rich corrosion products appeared to be the mechanism of corrosion for all the alloys tested. Inconel 600 and Inconel 690 showed better corrosion resistance than Inconel 625 and alloy 800H in molten LiCl-KCl salt in argon atmosphere[8]. The molybdenum in Inconel 625 did not improve the corrosion resistance in molten salt in the presence of air. Grain boundary attack and formation of precipitates within the grain and along the grain boundaries were observed in the SEM images for alloy 800H.

Visual examination of various carbon samples after the immersion test in LiCl-KCl molten salt for 2000 h revealed the formation of pores on the surfaces of low density graphite (LDG) and HDG and uniform attack was observed on the entire surface of these samples (Fig. 4)[9]. The surface of as-received glassy carbon (GC) appeared as a shiny layer; however, after corrosion test the surface became marginally dull without any attack on the surface (Figs. 5(a) and 5(b)). The degradation of LDG and HDG in molten salt is attributed to surface porosity filling and diffusion of salt into graphite, thereby dislodging the atoms from the lattice. However, formation of intercalation compounds was not observed. On the other hand, there was no change in the appearance of pyrolytic graphite (PyG) after the immersion test (Figs. 5(c) and 5(d))[9,10]. The corrosion resistance of carbon materials to molten salt was found to follow the sequence: low density graphite < high density graphite < glassy carbon < pyrolytic graphite[9].

3.2. Development of ceramic coatings and evaluation of their corrosion behaviour:

Plasma sprayed YSZ coating on 316L SS did not show much attack, unlike uncoated 316L SS. Surface morphologies of the exposed and unexposed samples appeared similar (Figs. 6(a) and 6(b)); however, salt deposits rich in Fe could be seen on the sample exposed in molten salt for 1000 h.

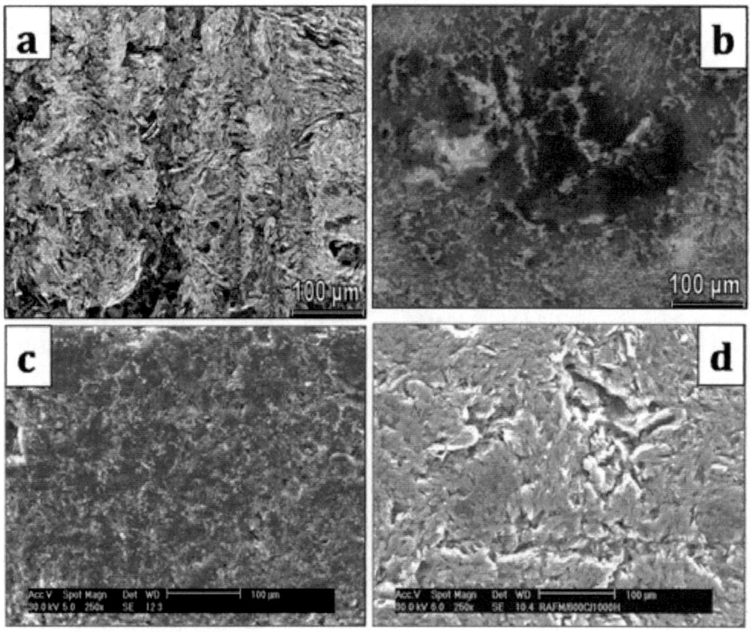

Fig. 4: Surface morphology of **(a)** as-received LDG **(b)** corrosion tested LDG **(c)** as-received HDG; **(d)** corrosion tested HDG, in molten LiCl–KCl salt for 2000 h

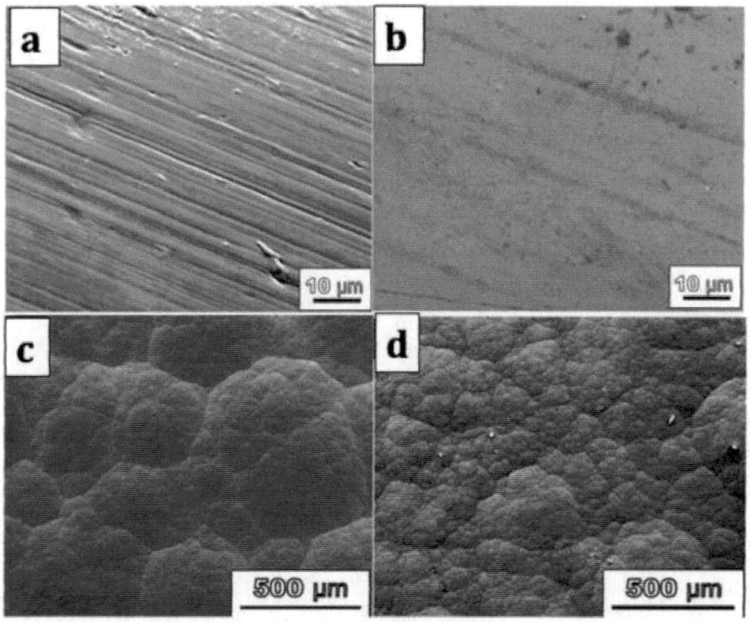

Fig. 5: Surface morphology of **(a)** as-received GC; **(b)** corrosion tested GC; **(c)** as-received PyG; **(d)** corrosion tested PyG in molten LiCl–KCl salt for 2000 h

Fig. 6: SEM images of YSZ coated 316L SS: **(a)** un-exposed and **(b)** exposed to molten LiCl-KCl salt for 1000 h

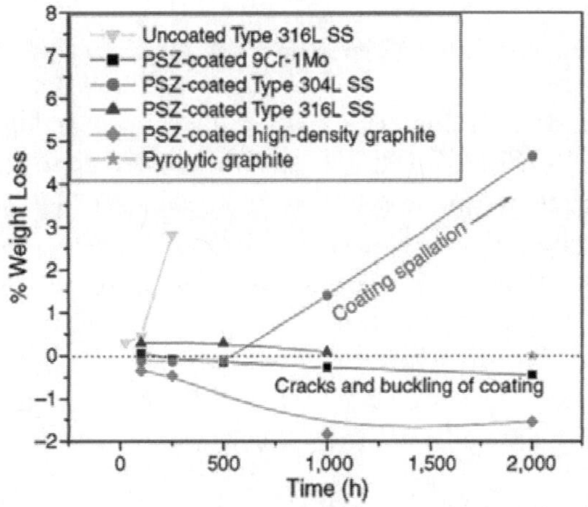

Fig. 7: Percentage weight loss of YSZ coated candidate materials exposed to molten salt at 600°C

The cross-section optical micrograph of unexposed and exposed YSZ coated 316L SS were more or less similar and did not reveal any penetration of the salt across the coating. Thus, YSZ coating on 316L SS exhibited better corrosion resistance in molten LiCl–KCl salt[6].

The percentage weight loss of YSZ coated structural materials exposed to molten LiCl-KCl salt at 600°C for various durations is shown in Fig. 7 [7]. Negligible weight loss observed for the YSZ coated HDG sample after exposure to molten salt for 2,000 h is evidence that YSZ coating performed well in molten LiCl-KCl salt. However, spallation of the coating was observed in few of the exposed HDG samples. The YSZ coated 9Cr-1Mo steel samples tested up to 2,000 h showed a minor increase in weight and the coating exhibited significant cracks and buckling. Though spallation of YSZ coating on 304L SS led to an increased weight loss, the coated specimens exhibited better corrosion resistance in molten LiCl-KCl salt than that of uncoated specimens.

The corrosion resistance of YSZ coating (Fig. 8) developed on HDG samples was evaluated in molten salt medium[11]. The coated samples, tested in LiCl-KCl molten salt for 250, 1,000 and 2,000 h showed a marginal weight gain owing to the presence of salt deposits. The corrosion attack on the uncoated HDG sample was more severe when compared to YSZ coated specimens.

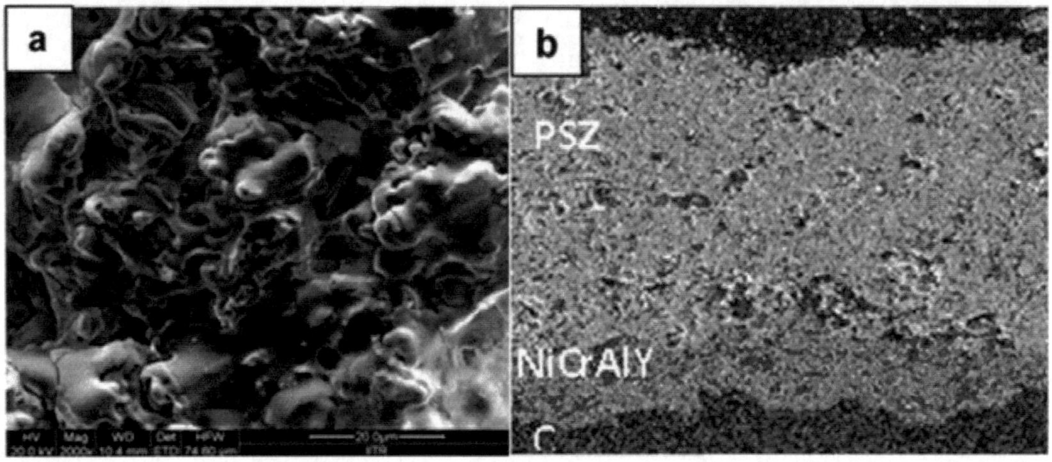

Fig. 8: As-received YSZ coating on HDG substrate: **(a)** top surface and **(b)** cross-section image

SEM micrographs of unexposed and exposed HDG samples to molten LiCl-KCl salt at 600°C revealed that the attack by molten salt increased with immersion time. There was no change in the surface morphology of YSZ coating after 2,000 h immersion test. Nevertheless, all the samples exposed to molten salt were found to form a dense outer layer and deposition of salt particles over the surface[11].

Coatings of YSZ with NiCrAlY bond coat and alumina-40% titania (A40T) with Cr_3C_2–NiCr bond coat were deposited on HDG and their performance was evaluated in uranium melting experiments. YSZ coating was intact after the uranium melting, but A40T coating peeled off completely along with bond coat. It was observed that A40T coating exhibited poor adhesion on HDG, but the A40T coating deposited without bond coat showed better bond strength. YSZ coating maintained its integrity without any phase change in the coating after uranium melting at 1350°C and no reaction with uranium was observed, as inferred from XRD and Raman studies. The surface morphologies of YSZ coatings after uranium melting were similar to those of as-sprayed coating; however, micro cracks were generated due to stresses[12].

Nanocrystalline TiN, ZrN and Ti–Si–N coatings were deposited on HDG disc samples and crucibles by DC/RF magnetron sputtering. Diffraction peaks corresponding to crystalline TiN phase were observed in Ti–Si–N coated sample and there was no indication for the formation of either crystalline Si_3N_4 or titanium silicide phases. Cross section SEM micrographs confirmed adequate adhesion of the coatings. Micro-scratch test showed adhesive failure, but no significant detachment of coating was observed after a blank run and in uranium melting even with a very high heating rate of 85°C/min[13].

Visual examination of the yttria coated HDG samples with NiCrAlY bond coat (Fig. 9(a)) after thermal cycling experiments revealed the failure of the coating from one edge of the sample during 36th and 20th cycles at 1000 and 1100°C, respectively. Elemental analysis confirmed that delamination of the coating occurred at the interface between the bond coat and top coat; but not completely from the HDG substrate. Non-uniform or insufficient thickness of bond coat could have resulted in the crack formation, as neither yttria top coat nor NiCrAlY bond coat powder undergoes any phase change at 1100°C (Fig. 9(b)).

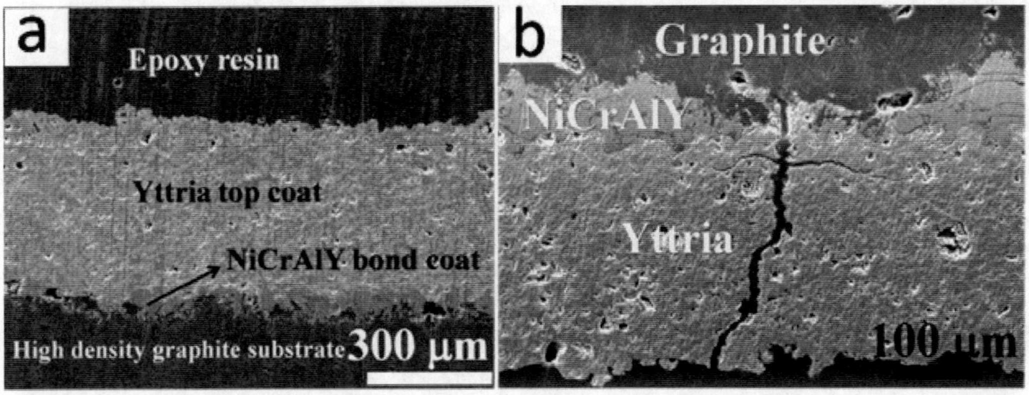

Fig. 9: Cross section image of yttria coating with NiCrAlY bond coat on HDG: **(a)** before **(b)** and after thermal cycling

Analysis of XRD patterns and Laser Raman spectra of yttria coated HDG samples confirmed the presence of only cubic yttria phase in both as-received and thermal cycled samples, implying thermal expansion mismatch among the coatings and the substrate had resulted in the failure of coating[14].

Uranium melting experiments revealed the compatibility of yttria coating with molten uranium for three melting cycles without coating spallation. Micro-cracks generated on the surface of the coating could be owing to either mismatch in thermal expansion coefficients between the top and bond coats or the tendency of NiCrAlY bond coat to undergo melting beyond 1300°C. Since NiCrAlY bond coat failed during the uranium melting and thermal cycling studies, for the purpose of improving the coating quality, yttria coating without any bond coat was developed on HDG crucibles for molten metal applications. Plasma sprayed yttria coating without bond coat performed well for more than 40 copper melting cycles at 1150°C and for 5 copper melting cycles at 1400°C. Thermal cycling studies of yttria coated HDG disc without bond coat performed well for 244 cycles at 1000°C and thermal cycling studies of at higher temperatures are in progress. Yttria coated HDG without bond coat also performed well for 8 uranium melting cycles (Fig. 10(a)) and further runs are in progress.

Yttria coating on HDG without bond coat failed after 5th thermal cycle itself at 1400°C and above. Thermal mismatch assisted crack initiation at the yttria free surface propagated to the constrained interface, which eventually caused top coat delamination. X-ray diffraction pattern and EDS spectra did not reveal any yttrium carbide phase at the interface of the failed coating after thermal cycling. For further improvement of yttria coating on HDG for uranium melting applications, yttria coating on HDG with a highly stable interlayer coating is being developed. Literature survey and thermodynamic calculations revealed that carbides of Si, Zr, Nb and Ta are very stable and compatible with graphite. Recently, SiC interlayer by pack cementation process was attempted. A dense uniform SiC layer was developed on HDG disc samples by pack cementation process using optimized temperature and duration, in a powder pack consisting of SiO_2, SiC and carbon. A dense layer of yttria was deposited as top coat by APS process over the SiC bond coat. The coating failed during 22nd thermal cycle at 1400°C (Fig. 10(b)) and during 12th cycle at 1450°C[15].

Fig. 10: (a) Photograph of HDG crucible coated with yttria without bond coat and molten uranium **(b)** SEM image of yttria coating on HDG with SiC bond coat after thermal cycling

4.0. Conclusions

Surface engineering plays a14n important role in the corrosion behaviour of various structural materials for high temperature applications. The corrosion resistance of several structural materials was evaluated in LiCl-KCl molten salt at high temperature. Ceramic coatings with and without bond coats were developed on the structural materials and the performance of these coatings in corrosive molten salt as well as in molten uranium and the durability of the coatings during thermal cycling were investigated. These studies revealed that HDG crucible with YSZ coating is suitable for molten salt purification system, 9Cr-1Mo steel with or without YSZ coating can serve as the candidate material for electrorefining vessel and a suitable carbide interlaid yttria coated HDG crucible is a promising material for both cathode processor and injection casting vessels for the pyrochemical reprocessing of spent metallic fuels of future sodium cooled fast breeder reactors.

5.0. References

1. Baldev Raj: Energy Procedia, 2011, vol. 7, p. 186.
2. Nagarajan, K.; Reddy, B.P.; Ghosh, S.; Ravisankar, G.; Mohandas, K.S.; Kamachi Mudali, U.; Kutty, K.V.G.; Kasiviswanathan, K.V.; Anand Babu, C.; Kalyanasundaram, P.; Vasudeva Rao, P.R.; Baldev Raj: Energy Procedia, 2011, vol. 7, p. 431.
3. Laidler, J.J.; Battles, J.E.; Miller, W.E.; Ackerman, J.P.; Carls, E.L.; Progress in Nuclear Energy, 1997, vol. 31, p. 131.
4. Westphal, B.R.; Marsden, K.C.; Price, J.C.; Laug, D.V.; Nucl. Eng. Technol., 2008, vol. 40, p. 163.
5. Kamachi Mudali, U.; Ravi Shankar, A.; Mallika, C.; Thyagarajan, K.; Ningshen, S.; Reddy, B.P.; Nagarajan, K.;: Materials and Coating Technologies for Pyrochemical Reprocessing of Spent Metallic Fuel from Fast Breeder Reactors, Internal Report: IGC/MMG/CSTG/2011, 2011, p. 1.
6. Ravi Shankar, A.; Kamachi Mudali, U.; Mater. and Corros., 2008, vol. 59, p. 878.
7. Ravi Shankar, A.; Thyagarajan, K.; Kamachi Mudali, U.; Corros., 2013, vol. 69, p. 655.
8. Ravi Shankar, A.; Kanagasundar, A.; Kamachi Mudali, U.; Corros., 2013, vol. 69, p. 48.

9. Jagadeesh Sure; Ravi Shankar, A.; Ramya, S.; Mallika, C.; Kamachi Mudali, U. Carbon, 2014, vol. 67, p. 643.
10. Jagadeesh Sure: Ravi Shankar, A.; Kamachi Mudali, U.; Nowicki, A.; Raj, B.; Surf. Eng., 2013, vol. 29, p. 28.
11. Jagadeesh Sure; Ravi Shankar, A.; Ramya, S.; Kamachi Mudali, U.; Cer. Int., 2012, vol. 38, p. 2803.
12. Jagadeesh Sure; Jagadeeswara Rao, Ch.; Venkatesh, P.; Prabhakara Reddy, B.; Mallika, C.; Kamachi Mudali, U.; Cer. Int., 2014, vol. 40, p. 6509.
13. Ravi Shankar, A.; Prabhakara Reddy, B.; Vipin Chawla; Jeya Priyanga, M.; Ramesh Chandra; Kamachi Mudali, U.; Surf. Coat. Technol., 2010, vol. 204, p. 3214.
14. Jagadeeswara Rao, Ch.; Ravi Shankar, A.; Mallika, C.; Kamachi Mudali, U.; Cer. Int., 2015, vol. 41, p. 3128.
15. Vetrivendan, E.; Thirunavukarasu, H.; Jagadeeswara Rao, Ch.; Mallika, C.; Kamachi Mudali, U.; Development of silicon carbide interlayer coating on graphite by pack cementation for deposition of yttria by plasma spray process, Surf. Coat. Technol. (Communicated).

Performance of Aluminum Diffusion Coating on Ferritic-Martensitic Stainless Steel P91 against KCl-Induced High Temperature Corrosion

S. Kiamehr*,
Technical University of Denmark (DTU),
Department of Mechanical Engineering,
Produktionstorvet, Building 425, 2800 Kgs. Lyngby, Denmark.

T.N. Lomholt,
FORCE Technology, Park Allé 345, 2605 Brøndby , Denmark.

K.V. Dahl, M.A. J. Somers
Technical University of Denmark (DTU),
Department of Mechanical Engineering,
Produktionstorvet, Building 425, 2800 Kgs. Lyngby, Denmark.
*Email: sabag@mek.dtu.dk & saeedbqs@yahoo.com

Abstract

Pack cementation was used to produce a Fe_2Al_5 coating on the ferritic-martensitic stainless steel P91. The performance of the coating against KCl-induced high temperature corrosion was evaluated by exposing the samples at 600°C for 168h in static lab air with and without a KCl deposit. Chemical and phase composition of the corrosion products were studied with Scanning Electron Microscopy (SEM), Energy Dispersive X-Ray Spectroscopy (EDS) and X-Ray Diffractometry (XRD). The coating showed excellent performance in the KCl-free exposure, but its performance in the presence of KCl was not satisfactory. The dominant mode of attack on the coating was selective aluminum removal. At some locations the coating was completely corroded.

1.0. Introduction

Effective utilization of biomass as a potentially CO_2-neutral fuel has been hampered due to the high temperature corrosion of heat exchangers. This is induced by the presence of potassium chloride (KCl)-rich deposits in an oxygen-containing environment. The current investigation aims at evaluating the performance of a diffusion coating against KCl-induced high temperature corrosion at a target temperature of 600°C. Choice of this temperature was based on the fact that, in the state of the art power plants, currently metal temperature is kept below 560°C to keep the corrosion rate at an acceptable level[1–2]. Increasing this temperature to 600°C significantly improves the electrical efficiency of the power plants utilizing biomass as the fuel. Unfortunately field tests involving several chromia-forming high temperature alloys have not shown satisfactory results[1–4]. Therefore evaluation of coatings relying on other oxide-forming elements such as aluminum appears to be an attractive option. In a previous work it was indicated that there was no significant reactivity between Al_2O_3 and KCl at 650°C[5]. Consequently aluminum enrichment at the surface of currently used stainless steels is expected to increase the corrosion resistance.

According to the Fe-Al phase diagram (Fig. 1) several intermetallic phases are expected to form by the interdiffusion between Fe and Al. In this study the focus is on the corrosion resistance of the Fe_2Al_5 phase, since it has been reported that the diffusion of the oxide-forming element in this phase is relatively fast[6]. Hence it is expected that a protective oxide can be established rapidly and maintained by fast replenishing of Al.

2.0. Experimental Procedure

A conventional boiler material, ferritic-martensitic stainless steel P91 (Fe-9Cr-1Mo-0.5Si-0.2V wt%), was chosen as the substrate alloy to be aluminized. Samples were coupons of ~ 20x8x0.3 mm. Prior to coating, the samples were ground with 500grit SiC paper and subsequently cleaned with ethanol. Pack cementation was performed in a mixture consisting of 10%Al, 6%$AlCl_3$ and 84%Al_2O_3 (in wt.%). The powder mixture and samples were packed in a closed alumina container. The pack was heated to 650°C and held for 6h under continuous flow of argon. After this period the furnace was shut down and the pack was allowed to cool down inside the furnace.

After coating, samples were cleaned in ethanol in an ultrasonic bath. One sample was partly (2/3 of the length) covered with a ~1mm thick layer of KCl powder and exposed for 168h at 600°C. For simplicity static lab air was chosen as the oxygen-containing environment. Under similar conditions a salt-free exposure was also performed for comparison. The as-coated and exposed samples were investigated with XRD (Bruker D8 Discover, CrK_α radiation (λ=2.2897Å)). Thereafter cross sections of the samples were prepared by embedding in epoxy and grinding/polishing using ethanol as lubricant to prevent the dissolution of the water-soluble compounds. The cross sections were studied with a SEM (FEI Inspect S) equipped with an EDS detector (Oxford Instruments). Imaging was performed in Back-Scatter Electron (BSE) mode.

3.0. Results

The microstructure of the as-produced coating is shown in Fig. 2. The deposited Fe_2Al_5 had a non-uniform thickness (~20-55μm) and often showed cracks perpendicular to the alloy/coating interface. Composition mapping with EDS (see Fig. 2) revealed that a chromium-rich phase, in form of narrow bands, was present parallel to the growth direction of the Fe_2Al_5. This phase extended from the alloy/coating interface towards the surface.

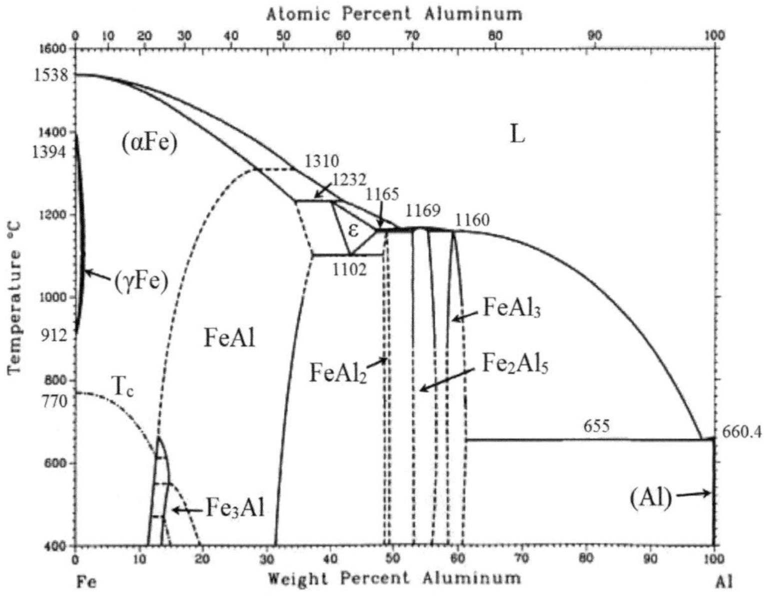

Fig. 1: Fe-Al binary phase diagram[7]

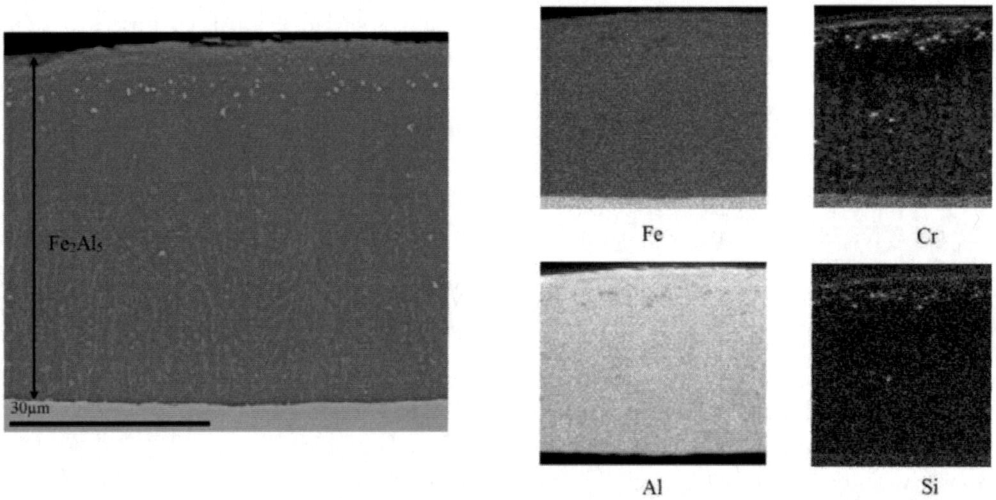

Fig. 2: As-coated Fe_2Al_5 on P91 and the distribution of the major alloying elements throughout the coating.

In addition, chromium and silicon- rich particles were detected in the outer part of the coating.

Salt-free exposure did not lead to significant surface degradation. An interdiffusion zone had formed in between the Fe_2Al_5 and the substrate alloy (Fig. 3). This zone was enriched in Al and depleted in Fe and Cr compared to the substrate.

Unlike the passive behavior observed for the salt-free exposure, the presence of KCl caused severe attack. The generally observed corrosion morphology was selective removal of aluminum and formation of a voluminous and porous aluminum-rich corrosion product on top of the coating (Fig. 4(a)).

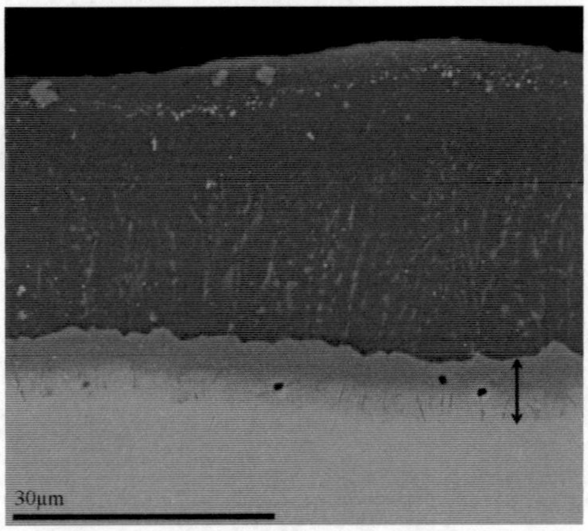

Fig. 3: Microstructure of the alloy and the coating after salt-free exposure. The interdiffusion zone is marked by the arrow.

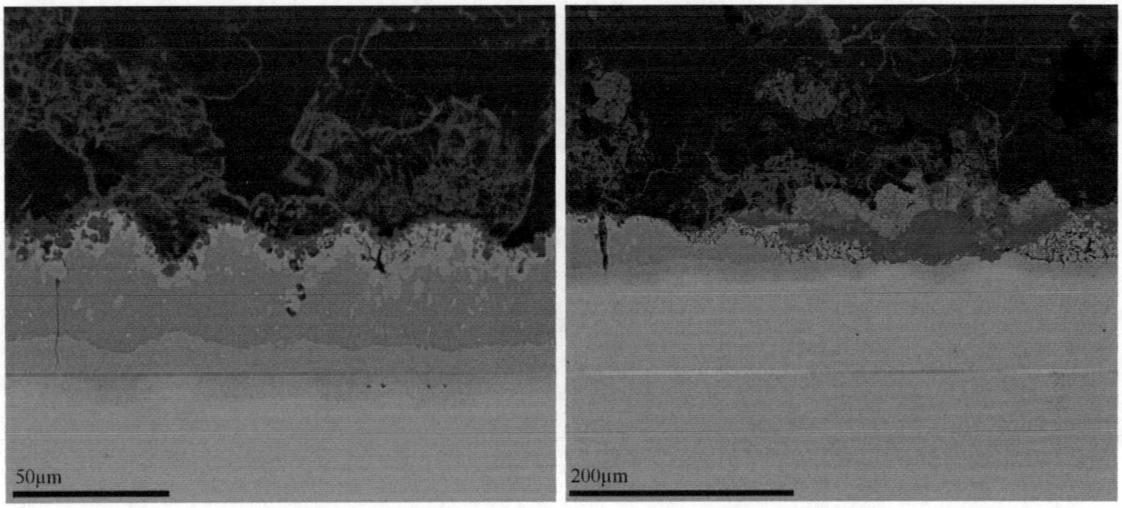

(a) (b)

Fig. 4: Corrosion product morphologies due to exposure under a KCl deposit in static lab air **(a)** general corrosion **(b)** local attack in form of through-thickness corrosion of the coating.

In a few cases Fe_2Al_5 was completely corroded and a thick double layer oxide had formed in addition to the outermost highly porous oxide (see Fig. 4(b)). In both cases potassium was found throughout the corrosion products (Figs. 5 and 6). At the interface between the metal and the residual coating an interdiffusion layer had formed similar to the salt-free exposure.

Fig. 7 shows the results of phase analysis with XRD. The as-coated layer consists of Fe_2Al_5 with a small amount of Cr_3Si consistent with the mapping in Fig. 2. The salt-free exposure does not lead to a significant change compared to the as-deposited coating. The diffractogram after the exposure to KCl shows the presence of α-Al_2O_3. No other oxide and/or potassium containing compound could be unambiguously detected. In addition, increasing the incidence angle to 5° (not shown in Fig. 7) revealed that a small amount of $Fe_{1-x}Al$ formed as well.

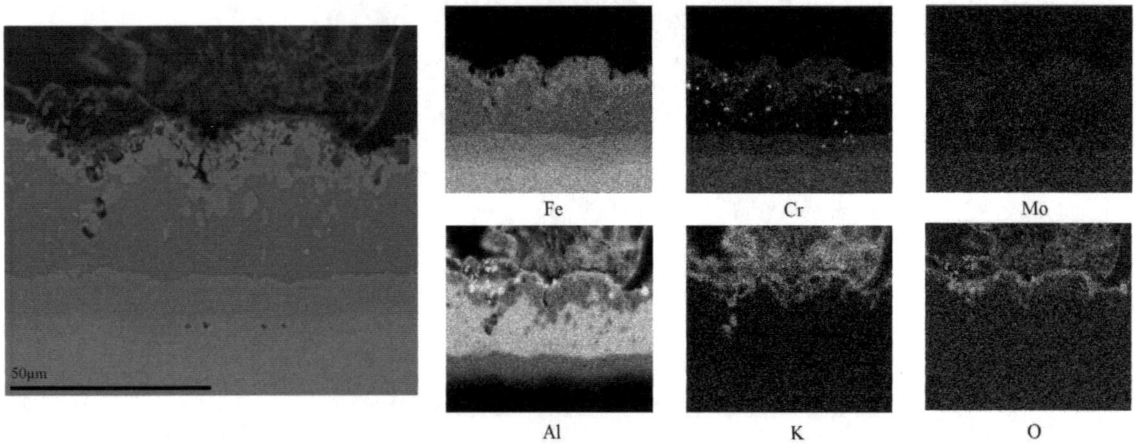

Fig. 5: Distribution of elements throughout the general corrosion morphology. The chlorine map is not included because the epoxy also contains chlorin

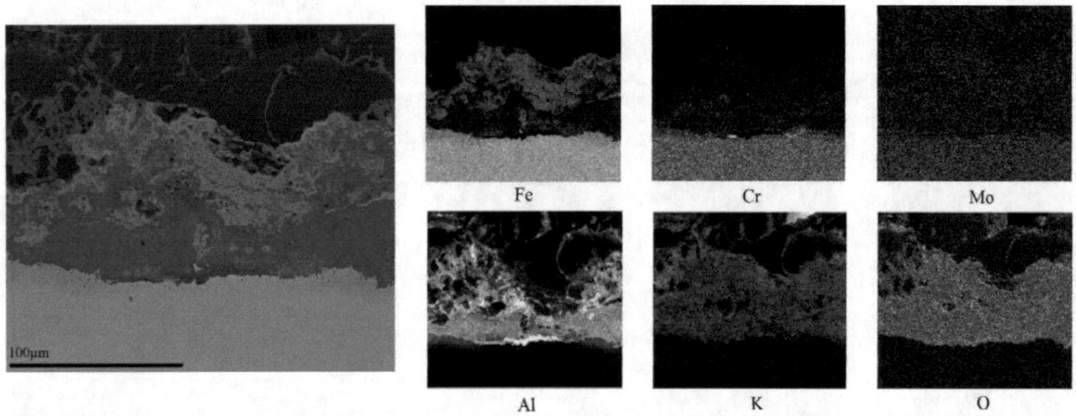

Fig. 6: Distribution of elements throughout the corrosion product morphology when the coating is completely corroded. The chlorine map is not included because the epoxy also contains chlorine

4.0. Discussion

Considering the diffractogram for the sample after salt-free exposure it is indicated that no oxide peaks developed and that all the peaks matched the diffractogram of the sample before the exposure. This suggests that an aluminum-rich protective oxide has formed that is too thin to be clearly identified with GI-XRD. This is consistent with the sample's microstructure (Fig. 3) showing the absence of any fast-growing oxide.

The presence of KCl has a remarkable effect on Fe_2Al_5 and the entire surface of the coating was affected. Fig. 4 and the mapping in Fig. 5 show that aluminum has been selectively removed from the coating's surface and consequently iron and chromium are enriched. In addition, the removed aluminum can be found outside the original coating's surface matching the maps of oxygen and potassium. This is consistent with the α-Al_2O_3 peaks found in the diffractogram shown in Fig. 7. However, the presence of a crystalline potassium-aluminum-oxygen compound could not be unambiguously confirmed by XRD.

The depth of aluminum removal as well as the spongy and voluminous nature of the corrosion product (Fig. 4(a)) suggests that the mechanism of depletion is volatilization of reactive alloying elements by chlorine.

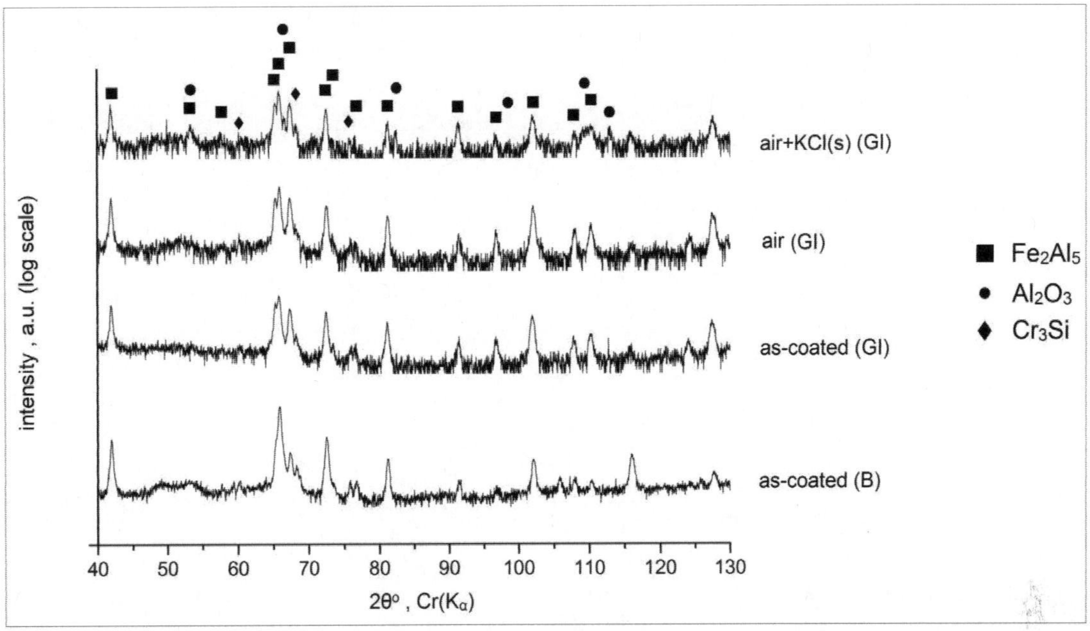

Fig. 7: X-ray diffractograms before and after exposures. (B) denotes measurement in Bragg-Brentano configuration. Other measurements were performed in grazing-incidence (GI, 2°) configuration

The same mechanism has been reported for field exposures of commercial alloys.[1-2] However, it is unclear whether the failure is due to the Fe_2Al_5 phase itself or whether it is triggered by an interaction with precipitate phases within the coating. Vokal et al.[8] attribute the KCl-induced attack on aluminide phases to a preferential chlorine attack on Cr-rich phases, α-Cr and/or $Cr_{23}C_6$, lying along grain boundaries. Metsäjoki et al.[9] report that the Cr-rich phases within a Fe_2Al_5 coating on P91 are Cr_3Si and Cr_2Al. In the present study only the presence of Cr_3Si could be confirmed with XRD. In addition, no clear sign of chromium involvement could be observed and it was aluminum, which was often associated with potassium.

An important observation was that the through-thickness corrosion of the coating only was observed at a few locations. It appears that if the depleted area maintains a sufficient aluminum content after an initial depletion, it can re-passivate. EDS spot analysis on several aluminum-depleted areas gave values around 51at.% for the residual aluminum. This is consistent with the detection of $Fe_{1-x}Al$ when the GI-XRD was performed at 5° and can describe why the corrosion does not continue on some locations. If the residual aluminum content after an initial de-aluminization is high enough to stabilize $Fe_{1-x}Al$, it seems that the alloy can re-passivate. Description of the corrosion morphology shown in Figure 4b is not straight-forward. However, one can speculate a case where the chlorine-induced de-aluminization of Fe_2Al_5 is locally so severe that the residual aluminum content is less than what is needed to stabilize the $Fe_{1-x}Al$. Accordingly the alloy cannot re-passivate after the initial stages of the attack and eventually a double layer oxide rich in iron at the top and aluminum at the bottom forms underneath the earlier corrosion products. In such case, as the entire coating is converted to oxides, potassium can be found all over the corrosion products. Such description is tentative and proving it requires time-resolved exposures.

5.0. Conclusions

The following conclusions can be drawn from this study:

- Pack cementation process was employed to produce Fe_2Al_5 on the ferritic-martensitic stainless steel P91. The coating showed cracks and non-uniformities in the thickness. However, it exhibited excellent performance against air-oxidation at 600°C.
- Fe_2Al_5 coated on P91 was heavily attacked when KCl was present. The attack was generally in the form of selective aluminum removal. On large parts of the surface the attack does not seem to have continued. At a few locations, the coating suffered from through-thickness corrosion.
- The (local) through-thickness corrosion of the coating and the brittle nature of the Fe_2Al_5 formed on P91 make the coating unsuitable for actual application in biomass fired plants. The ability of the $Fe_{1-x}Al$ phase to stop or delay attacks is of interest for future studies.

6.0. Acknowledgments

This work was performed within GREEN framework financed by the Danish Council for Strategic Research. John C. Troelsen, Flemming B. Grumsen and Peter J.S. Westermann are acknowledged for their help and technical support.

7.0. References

1. Montgomery, M.; Karlsson, A.; Mater. Corros., 1999, vol. 50, p. 579.
2. Montgomery, M.; Karlsson, A.; Larsen, O.H.; Materials and Corroion., 2002, vol. 53(2), p. 121.
3. Montgomery, M.; Biede, O.; Larsen, O.H.; Mater. Sci. Forum, 2006, vol. 522–523, p. 523.
4. Henderson, P.; Davis, C.; Montgomery, M.; VGB powertech, 2005, vol. 6, p. 53.
5. Kiamehr, S.; Dahl, K.V.; Lomholt, T.N.; Christiansen, T.L.; Somers, M.A.J.; Int. Symposium on High Temp. Oxid. Corros., 2014.
6. Sakidja, R.; Perepezko, J.H.; Calhoun, P.; Oxidation of Metals., 2014, vol. 81(1–2), p. 167.
7. Metals Handbook, Alloy Phase Diagrams. ASM International, 1992, vol. 3.
8. Vokal, V.; Rohr, V.; Pomeroy, M.J.; Schutze, M.; Mater. Corros., 2008, vol. 59(5), p. 374.
9. Metsajoki, J.; Huttunen-Saarivirta, E.; Lepisto, T.; Fuel, 2014, vol. 133, p. 173.

Influence of Shot-Peening on the High Temperature Oxidation of Zr and Ti Plates

V. Optasanu*, A.Kanjer, T. Montesin
ICB UMR 6303 CNRS - Université de Bourgogne,
9 Av Alain Savary, BP. 47870,
21078 Dijon Cedex, France .
*Email: virgil.optasanu@u-bourgogne.fr

Abstract

The influence of the shot-peening on the oxidation rate of Zr is presented. After shot-peening, the samples are oxidized at temperatures between 350°C and 650°C for 3 days. It is shown that the oxidation rate substantially decreases by shot-peening at high temperature. At lower temperatures (T<500°C) the effect is opposite. At high temperature the main diffusion mechanism is the interstitial lattice diffusion of O^{2-} ions within the metal. The slowing of the oxidation rate is partially explained by the mechanical stress distribution and the differences in the oxide morphology. A thermodynamic model using strong diffusion/stress coupling applicable to the bulk diffusion is used to explain the influence of the residual stress on diffusion. At lower temperatures, the oxidation is governed by grain-boundaries and dislocations diffusion and then the high dislocation density produced by shot peening explains the increase of the oxidation rate. First investigations on high temperature oxidation of Ti show also a decrease of the oxidation rate with the shot-peening.

1.0. Introduction

The Zirconium and its alloys are used as cladding element in the nuclear industry due to its transparency to neutrons and good oxidation resistance at moderate temperatures. The study of the high temperature oxidation of this metal concerns mainly the security of the nuclear power plants. The Titanium presents huge industrial interest and is increasingly used in high temperature aeronautics and aerospace domains.

The shot peening is used to introduce high residual stress by producing strong plastic strains and hardening on the superficial layers of the samples. Well known as a process used for fatigue life improvement, the influence of shot peening on high temperature oxidation is much less investigated[1–3]. It was experimentally remarked that at high temperatures, the strong hardening of the superficial layer produces a slowing down of the oxidation rate.

In our previous works we shown that shot-peened Zr samples have oxidation rates sensibly lower than the original material for short[1] (30 min), or longer oxidation times[3] (3 days) at 650°C. The Zirconium is generally not used uncoated at high temperature. The knowledge of the high temperature oxidation of this metal is important for security reason as it is used as first security barrier in nuclear reactors. Uncoated Titanium is used in aeronautic and aerospace engineering at high temperatures (jet engine deflector parts).

The high temperature oxidation (>450°C) of Zirconium is governed by the bulk diffusion of O^{2-} ions into the metal lattice while at medium temperatures the diffusion process mainly occurs in grain boundaries[4]. At high temperature the diffusion coefficient has sufficiently large values to give large diffusion zones of several tens of microns. The insertion of large quantities of oxygen into the zirconium lattice (up to 29% atomic) tends to expand the metal lattice and then gives important compression stress values in the plane parallel to the oxide/metal interface.

The diffusion process is closely influenced by the mechanical stress[5–7]. As shown in the literature[8,9] the apparent diffusion coefficient depends on the mechanical stress as well on the stress gradients and can influence the oxidation based on anionic diffusion of the oxygen into the metal lattice[1,10,11]. By using pure thermodynamic developments the first Fick's law can be modified as follows[1]:

$$\vec{J} = -D_0\left[\left(1+\frac{M_0}{RT}\eta_{ii}\sigma_{ii}\right)\vec{\nabla}c - \frac{M_0}{RT}c\vec{\nabla}\left(\eta_{ii}\sigma_{ii}\right)\right], \quad (1)$$

where $\vec{J}$ is the flux of oxygen into the metal, M_0 is the molar mass of the metal, c is the oxygen concentration, σ_{ij} is the stress tensor and $\eta_{ij} = \frac{\partial \varepsilon_{chem}}{\partial c}$ is the chemical expansion coefficient[2]. For example, in the Zr case, the average value of the chemical expansion coefficient is $\eta = 3.55\ 10^{-5}\text{m}^3\ \text{kg}^{-1}$.

The Eq. (1) is purely thermodynamic and do not contains any phenomenological parameter. It shows two driving forces one governed by the concentration gradient but influenced by the mechanical stress, the second governed by the stress gradients.

For the physical interpretation of the stress influence on the diffusion law we will consider in the following explanations the case of a material that expands when oxygen diffuses within its crystallographic lattice (the chemical expansion coefficient, $\eta > 0$).

The influence of the stress on the first term $-D_0\left(1+\frac{M_0}{RT}\eta_{ii}\sigma_{ii}\right)\vec{\nabla}c$ can be illustrated as a flux increase in tensioned regions of the material. We notice that the presence of a compressive stress decreases the effective diffusion coefficient.

The influence of the second term $D_0\frac{M_0}{RT}c\vec{\nabla}\left(\eta_{ii}\sigma_{ii}\right)$ can be illustrated like a mechanical pump: the oxygen migrates from the more compressed to less compressed regions. We also notice that the sign of this term is opposite to the concentration gradient term (Fick's term).

A compression stress decreases the value of the apparent diffusion coefficient. As an example, a triaxial compression stress of about 400 MPa applied on Zr decreases the effective diffusion coefficient of about 30% at 650°C. While the stress value is limited by the mechanical properties of the material, the stress gradient can reach very strong values if the variation of the stress is made within a very short depth. The term containing the stress gradient can then be very influent and can act in some cases as an up-wind driving force for the dissolved oxygen, by opposing the "natural" flow from "rich" to "poor" concentration. The stress taken into account can be due to external forces or to thermal and mechanical processes. In our previous simulation works we have shown that the diffusion can be "manipulated" by the mechanical stress[1].

Works on the Surface Mechanical Attrition Treatment (SMAT)[2] show effects on high temperature oxidation of M5 zirconium alloy noticed also a slowing down of the oxidation introduced by the nanocrystallization of the surface. We previously studied[3] the influence of several factors on the oxidation of Zr at 650°C: shot-peening duration, nature of the oxidant atmosphere, surface pollution and intensity of the stress values. We shown that the longer is the shot-peening treatment the bigger is the influence on the oxidation. The present work investigates the oxidation behavior of the shot peening plates in a range of temperatures between 350 °C and 650 °C.

2.0. Experimental Procedure

Commercially pure Zr and Ti plates (99.8% purity, 2 mm thick Zr plates, and 99.6% purity, 1 mm thick Ti plates, Goodfellow) have been used in this study. Rectangular samples 50 x 25 mm^2 were cut, annealed at 750°C for 2h under dynamic secondary vacuum in order to decrease the eventual post-manufacturing residual stress. The samples were then shot-peened for 5, 10, 20 or 30 min on each side, by alternate steps of 5 min/side in order to avoid a too strong Almen's deformation. The shot-peening is realized using 20 g of 2mm tungsten carbide balls and a sonotrode vibrating at 20 kHz with an amplitude of 12μm. In this study, the chosen shot-peening durations were 5, 10 and 20 min for Ti respectively 30 min for Zr.

After shot-peening, small samples of 10 x 10 mm^2 were cut and used for thermo-gravimetric analysis. Samples were subjected to oxidation at temperatures between 350°C and 650°C for few days. During the high temperature exposure the grains of the metal do not grow[1]. In contrast, the temperature is high enough to produce (at least partially at medium temperatures) the dislocation recovery and then substantial decrease the residual stresses. It is reported[12] that the total recovery is produced for less than 1h of annealing at 650°C for Zircalloy 2.

The residual stress after shot-peening has been measured by incremental hole-drilling method. Experimental details are presented elsewere[1]. Fig. 1 shows the profile of the residual stress in Zr plate in a direction parallel to the sample surface with respect to the depth. The stress curve has a particular shape characteristic to shot-peening: there is a first region on the first 150 μm with a negative gradient, followed by a zone with positive gradient. In the case of the oxidation, according to the Eq. (1), the first zone will contribute to decrease the oxygen flux into the metal bulk. Due to the positive stress gradient in the second zone (>150μm), the apparent diffusion coefficient value increases. The first zone slows down the diffusion, while the second zone accelerates it.

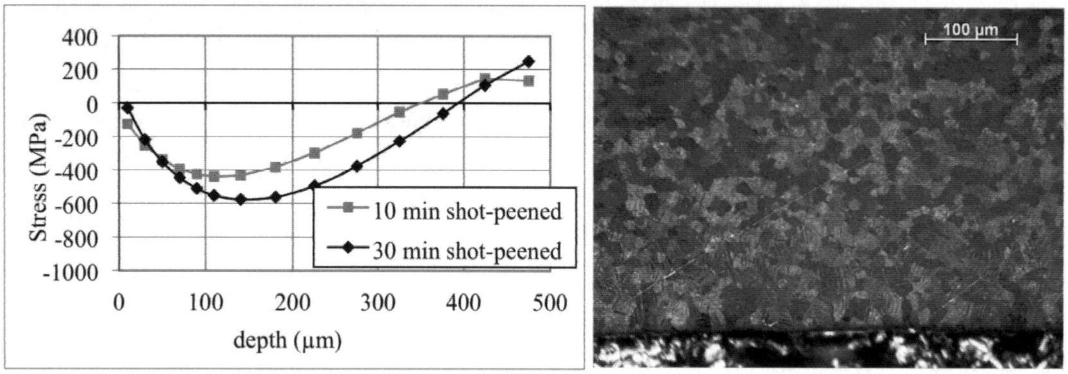

Fig. 1: Measured residual stress after shot-peening and micrographic image of the 10 min shot-peened sample

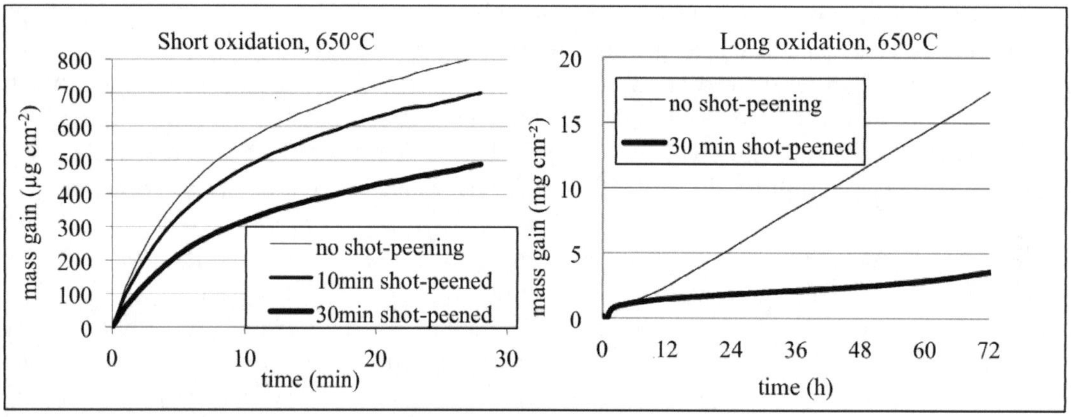

Fig. 2: Short and long oxidation durations for Zr plates, at 650°C

3.0. Results and Discussions

As shown in our previous works[1,3,13] on Zr plates, shot-peening can substantially decrease the high temperature oxidation rate at 650°C. Fig. 2 presents mass gain curves for short and long oxidation periods.

In the range of temperatures that we investigate the oxidation of Zr and Ti are anionic processes. The oxidation occurs by diffusion of the oxygen in the metal lattice followed by the transformation of the oxygen-richest zones into oxide. The concentration profile of the dissolved oxygen goes progressively from 29% atomic at the oxide/metal interface to zero in the pure-metal region. Fig. 3 presents the morphology of the oxide, interface and solid solution (of the oxygen in the metal), for pure Zr oxidized at 650°C for 3 days. One can remark that the oxide layer is cracked, with cracks parallel to the oxide/metal interface. The interface between metal and oxide is undulated. Also, between the pure metal and the oxide there is a zone with a variable amount of dissolved oxygen, as remarked on EDX measurements presented also Figure 3. In the pure metal zone, the apparent amount of oxygen is about 12%, but this is due to the pollution of the superficial layers by the atmospheric oxygen by the polishing of the sample during the sample preparation. Here, the length of diffusion of the oxygen into the metal lattice is about 15 μm.

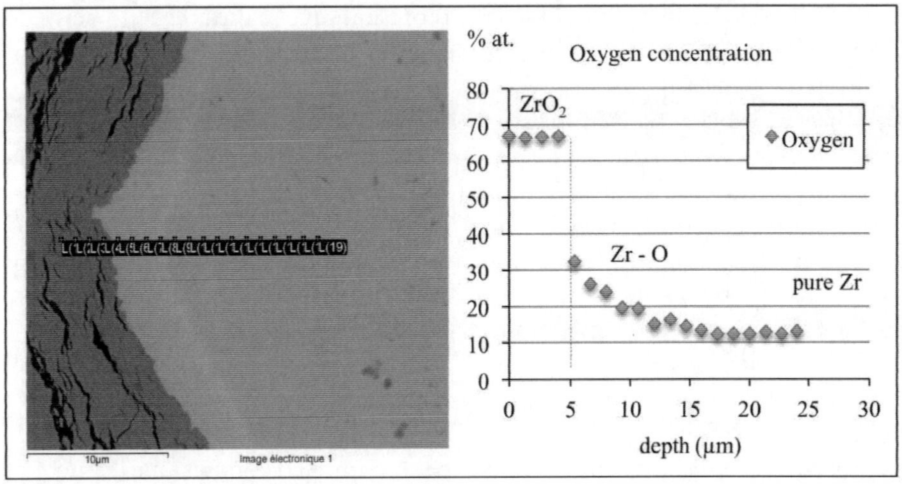

Fig. 3: Morphology of the oxide, interface and oxygen solid-solution zone (left). Corresponding EDX measurements of the oxygen concentration (right).

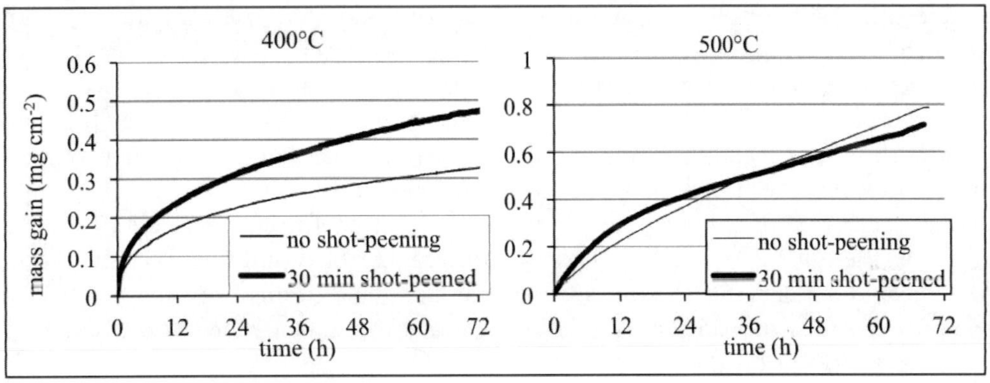

Fig. 4: Mass gain for Zr oxidized at 400°C and 500°C.

We published elsewhere[14] evidences that EDX measurements show length of diffusion shorter than other more accurate techniques like Nuclear Reaction Analysis.

The oxidation of the Zr was made at several temperatures: 350°C, 400°C, 450°C, 500°C, 650°C. Fig. 4 shows the influence of the shot peening on the oxidation for 400 and 500°C. For temperatures under 450°C the shot-peened Zr samples oxidize more than natural Zr samples. Under 450°C the mass gain is parabolic, which is the sign that the diffusion is the rate limiting process during the oxidation. For temperatures above 500°C the shot-peening slows down the oxidation rate. At 650°C (see Fig. 3) the mass gain at the beginning is parabolic, then linear, which is the sign that during the late stages of oxidation the protective un-cracked oxide layer thickness remains constant. This thickness is difficult to measure on micrographies because it is not constant. Moreover, the micrographic images are analysed after cooling at room temperature and some of them could appear during cooling. Also, the protectiveness of this layer is made by the non-connection of the cracks, which is impossible to quantify by micrographic techniques. At 650°C the mass gain is reduced by 80% - the rate of oxidation is divided by 5. The mechanical treatment of the samples produces a relative protection of the material at high temperature by slowing down the oxidation.

The different behavior at high and lower temperature can be explained by the lattice diffusion at high temperature while at lower temperatures the diffusion occurs mainly via grain boundaries and dislocations[4,15]. For Zr, under ~ 450°C the oxygen diffuses mainly by grain boundaries and dislocation, while above this temperature occurs mainly the lattice diffusion[4]. The density of dislocation is highly increased by shot-peening in the superficial layer. Thus, at low temperatures, the short-circuits of diffusion produce an increasing of the oxidation rate[15]. The oxide formed on shocked metal, can produce small oxide grains and subsequently many grain boundaries, which are the preferential way for the oxygen ions to reach the metal.

At high temperature the slowing can partially be explained, at the initial stages of the oxidation, by the mechanical stress acting as a driving force. Another factor that can play a role is the stabilization of the quadratic phase of zirconia by the high compression stress[16-19]and the small crystallites[17,20,21] size in the shocked area. The formation of zirconia layer is a complex phenomenon and is a combination of quadratic and monoclinic phases. The quadratic phase appears in high compression conditions and is the phase found at the interface between the metal and the oxide as recently shown by Raman spectroscopy[22]. There are experimental evidences[2] obtained by XRD techniques that shows that the oxide formed on the shocked layers are different from those formed on unshocked layers: on shot-peened layers the tetragonal/monoclinic ratio is bigger than in un-peened layers. The measure of the tetragonal/monoclinic ratio during the high temperature oxidation will be the object of our future works. First XRD phase measurements made on shot-peened and un-shot samples oxidized for 6h at 650°C and then cooled at room temperature, didn't show any statistically representative difference between phase composition. However, after the annealing of the metal, which occurs rapidly at high temperature (~ few hours at T> 500°C), only the early oxide can be morphologically different. Thus, the reason of this oxidation behaviour of shot-peened samples at high temperature is not obvious and has to be studied furthermore. This particular point will be the object of our future works.

Oxidation made on shot-peened then chemically etched Zr samples, where superficial layers of about 3μm thick have been removed, shows the same behavior in oxidation as the un-etched samples: more the sample is shot-peened lower is the mass gain. This shows that the eventual pollution of the surface by the WC balls is not responsible of the oxidation rate slowing down. However, etched samples (having a thinner shocked layer) oxidize more than un-etched specimens, which indicate that the highly shocked superficial layer has an important influence on the oxidation.

A coefficient evaluating the relative difference, between shot-peened and natural samples, in terms of mass gain during oxidation, can be defined as follows:

$$\delta = \frac{\Delta m_{sh-p} - \Delta m_0}{\Delta m_0},$$

where Δm_{sh-p} and Δm_0 are the mass gain for shot-peened and respectively natural samples. This coefficient is positive when the shot-peening accelerates the oxidation and negative when it slows it down. Bigger is the absolute value of δ, greater is the influence of the shot peening on the oxidation. Figure 5 presents values of the coefficient δ with the temperature for 30 min shot-peened Zr samples. A linear trend line shows that the influence of the shot-peening vanishes around 500°C. However at mass gain at 500°C, (see Fig. 4) shows an important difference between the shapes of the two mass-gain curves.

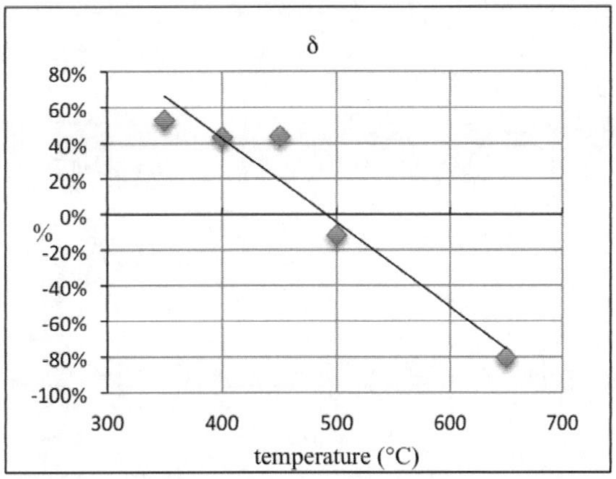

Fig. 5: Relative difference between mass gain of shot-peened and natural samples, δ coefficient

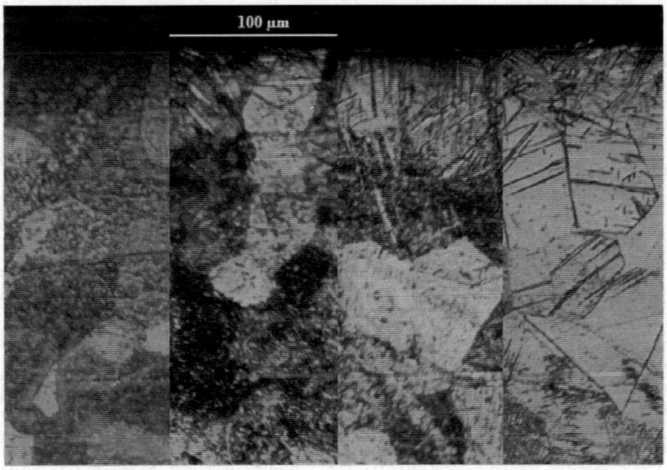

Fig. 6: Microstructure of Ti samples, from left to right: no shot-peening, 5, 10 and 20 min shot-peened

A longer oxidation time should give more difference between the final mass gains. Thus, for longer times of oxidation the neutral point should be about 500°C.

In the case of Ti the microstructure changes substantially with the shot-peening, as seen Fig. 6. Grain refinement up to 50 μm to the surface, as well twinning up to 200 μm can be seen on 20 min shot-peened samples. Figure 7 shows a mass gain for Ti at 650°C for different of shot-peening durations. As for Zr, the shot-peened Ti plates oxidize less than natural Ti plates. At this temperature, for 20h of coefficient $\delta = -47\%$ for 20 min of mechanical treatment, which is remarkable. Investigations at lower temperatures were not yet realized for Ti because of the good resistance to oxidation of this material, which would imply long oxidation tests necessity. Long exposure periods at lower temperatures are however interesting for industrial purposes and will be made in our future works.

4.0. Conclusion

In this work we studied the high temperature oxidation of shot-peened plates of Zr and Ti. We can conclude that:

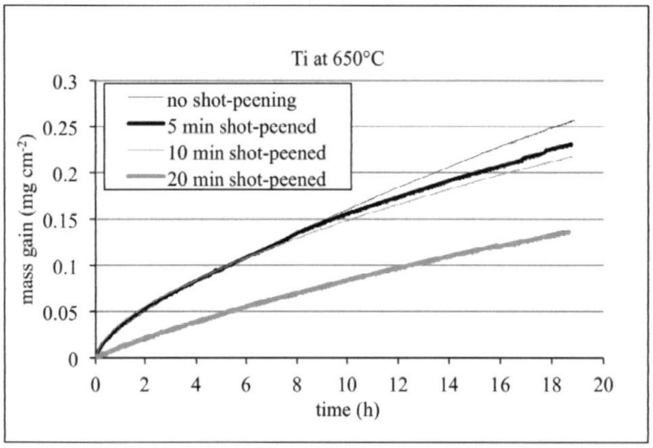

Fig. 7: Mass gain for shot-peened plates of Ti at 650°C.

- The shot-peening is a simple mechanical methods for life-time improvements of Zr and Ti exposed at high temperature oxidation.
- For Zr, at temperatures between 300°C and 500°C the shot-peening increases the oxidation rate.
- This acceleration happens in the domain of temperatures where the diffusion occurs preponderantly via grain boundaries and dislocations.
- For Zr, at T>500°C, the shot-peening produces a slowdown of the oxidation rate.
- This phenomenon is partially explained by the presence of the stress state that produces a mechanical driving force, and by the differences in the morphology of the oxide layer, which grows on a nanostructured substrate. However the entire mechanism of the slow down of the oxidation rate is not yet well understood and need further investigations.
- The eventual pollution of the surface during the shot-peening by the WC balls is not responsible for the slowing down of the oxidation rate because etched shot-peened samples still oxidize less than etched un-shot-peened specimens.
- Shot-peened Ti samples show a slow down of the oxidation rate with the shot-peening.

5.0. Acknowledgements

The authors would like to thank the LASMIS laboratory members (UTT, France) for the shot-peening treatment and stress measurements.

6.0. References

1. Raceanu, L.; 'Contribution à l'étude des interactions entre contraintes résiduelles et dissolution d'oxygène dans un solide déformable réactif', PhD Thesis, Université de Bourgogne, 2011.
2. Pannicaud, B.; Retraint, D.; Grosseau-Poussard, J.-L.; Li, L.; Guerain, M.; Goudeau, P.; Tamura, N.; Kunz, M.; Corros. Sci., 2012, vol. 60, p. 224.
3. Raceanu, L.; Optasanu, V.; Montesin, T.; Montay, G.; Francois, M.; Oxid. Metals, vol. 2013, vol. 79, p. 135.
4. Debuigne, J.; Lehr, P.; Revue de Métallurgie, 1963, vol. 60, p. 911.
5. Larché, F.; Cahn, J. W.; Acta Materialia, 1973, vol. 21(8), p. 1051.
6. Favergeon, J.; Montesin, T.; Bertrand, G.; Oxid. Metals, 2005, vol. 64, p. 25.

7. Creton, N.; Optasanu, V.; Montesin, T.; Garruchet, S.; Desgranges, L.; Defect and Diffusion Forum, 2009, vol. 289-292, p. 447.
8. Creton, N.; Optasanu, V.; Garruchet, S.; Raceanu, L.; Montesin, T.; Desgranges, L.; Dejardin, S.; Defect and Diffusion Forum, 2010, vol. 297-301, p. 519.
9. Raceanu, L.; Optasanu, V.; Montesin, T.; Creton, N.; Defect and Diffusion Forum, 2011, vol. 309-311, p. 149.
10. Optasanu, V.; Raceanu, L.; Montesin, T.; Defect and Diffusion Forum, 2012, vol. 323-325, p. 109.
11. Desgranges, L.; Plancher, H.; Gamaléri, M.; Micha, S.; Optasanu, V.; Raceanu, L.; Montesin, T.; Creton, N.; J. Nucl. Mater., 2010, vol. 402 (2-3), p. 167.
12. Lee, D.; J. Nucl. Mater., 1970, vol. 37, p. 159.
13. Optasanu, V.; Jacquinot, P.; Montesin, T.; Adv. Mater. Res., 2014, vol. 996, p. 912.
14. Optasanu, V.; Bourillot, E.; Vitry, P.; Plassard, C.; Beaurenaut, L.; Jacquinot, P.; Herbst, F.; Berger, P.; Lesniewska, E.; Montessin, T.; Nanoscale, 2014, vol. 6 (24), p. 14932.
15. Kofstad, P.; High Temperature Corrosion, Elsevier Applied Science Publishers Ltd., 1988.
16. Barberis, P.; J. of Nucl. Mater., 1995, vol. 226 (1), p. 34.
17. Bouvier, P.; Godlewski, J.; Lucazeau, G.; J. Nucl. Mater., 2002, vol. 300 (2), p. 118..
18. Godlewski, J.; 'Oxydation d'alliages de zirconium en vapeur d'eau : influence de la zircone tétragonale sur le mécanisme de croissance de l'oxyde', PhD Thesis, University of Technology of Compiegne 1990
19. Qin, W.; Nam, C.; Li, H.; Szpunar, J.; Acta Materialia, 2007, vol. 55 (5), p. 1695.
20. Bouvier, P.; Lucazeau, G.; J. Phys. Chem. of Solids, 2000, vol. 61 (4), p. 569.
21. Djurado, E.; Bouvier, P.; Lucazeau, G.; J. of Solid State Chem., 2000, vol. 149 (2), p. 399.
22. Kurpaska, L.; Favergeon, J.; Lahoche, L.; El-Marssi, M.; Grosseau Poussard, J-L.; Moulin, G.; Roelandt, J.-M.; J. Nucl. Mater., 2015 available online dx.doi.org/10.1016/j.jnucmat.2015.06.005

The use of Double Surface Treatments To Increase The Fatigue Life of a Quenched and Tempered High-Strength Steel

E. Segurado, J. Belzunce*
SIMUMECAMAT R&D group,
University of Oviedo, Campus Universitario, 33203 Gijón, Spain.
*Email: belzunce@uniovi.es; j.belzunce@itma.es

Abstract

Shot peening is widely used in the automotive and aeronautical industries as a mechanical surface treatment to improve the fatigue properties of metallic components. Generally, the effectiveness of shot peening processes is principally ensured by controlling the peening intensity. The higher the applied Almen intensity, the larger the surface-induced hardening and the deeper the residual compressive stress field, although surface damage in the form of tiny, shallow notches may occur. From this perspective, the use of double treatments is a good option to attain deep compressive residual stress profiles while minimizing the extent of surface defects.

Two different kinds of double treatments were studied in this research work: shot peening plus vibratory finishing and shot peening plus grit blasting. Alumina with a grit size below 0.1 mm and a low air pressure were employed in the latter case.

The fatigue life of the steel was greatly improved following the application of both double treatments. The best treatment procedures were likewise defined and the obtained results were justified by means of the residual stress profiles, roughness measurements and the location of fatigue crack initiation.

1.0. Introduction

Shot peening is widely used in the automative and aerospace industries as a mechanical surface treatment to improve the fatigue life of metallic components. It consists in projecting very hard, tiny, spherical balls onto the surface of the component to induce surface plastic deformation, the expansion of which is constrained by the adjacent deeper material. These surface treatments produce several effects (roughness, surface hardening, compressive residual stresses, but also surface defects) which, when properly controlled, can provide important enhancements, especially to ameliorate the fatigue behaviour of components subjected to dynamic loads [1–3].

On the other hand, different authors have demonstrated the possibility of increasing the fatigue life of a component using double treatments[4–6]. Basically, double treatments consist in first applying a high intensity peening treatment, the major effect of which is to produce a deep region submitted to high magnitude compressive stresses, followed by a second surface treatment to reduce the roughness induced in the first treatment and mitigate the damage produced on the surface. Polishing is the usual technique employed as a second treatment. The use of other, more effective types of second surface treatments, such as vibratory finishing and grit blasting, were explored in this study.

2.0. Methods

The study was performed on an F1272 steel (equivalent to AISI 4340). This is a 0.41% carbon steel alloyed with 1.9% nickel, 0.87% chromium and 0.24% molybdenum. The steel was supplied in bars with a nominal diameter of 16 mm. The bars were submitted to the following heat treatment: austenitization at 850°C for 45 min., following by water quenching and tempering at 200°C for 15 min. The heat-treated steel had a yield strength of 1596 MPa, an ultimate tensile strength of 2000 MPa, a tensile elongation of 11% and a reduction in area of 50%.

2.1. Surface treatments

Shot peening was performed by means of a direct compressed air machine using conditioned cut wire shots with rounded-off edges (CW, 670-730 HV, with an average diameter of 0.7 mm) under an Almen intensity of 19A and 100% coverage.

Two types of second treatments were subsequently applied over different treatment times: vibratory finishing and grit blasting. Vibratory finishing was applied by placing the samples in a Rösler vibratory finishing machine using abrasive bevelled ceramic cylinders with a length of 10 mm and a diameter of 3 mm (RP 03/10 ZS), along with a SiC polishing paste (RSP 508 S) and water. Different working times up to 24 hours were applied. Grit blasting, on the other hand, was applied by means of the same direct compressed air machine using Al2O3 powder with a grit size below 0.1 mm and an air pressure of 2 bars. Different treatment times were likewise applied in this case.

2.2. Characterization techniques

Roughness measurements, residual stress profiles and fatigue tests were used as characterization techniques following the single shot peening 19A treatment and also after the different double treatments. Surface roughness was determined on a roughness tester by means of the average, Ra, and the maximum roughness, Rmax. Five different roughness profiles were performed on each sample and the average results were reported.

Residual stress profiles were determined by X-ray diffraction and incremental layer removal by electropolishing using the $\sin^2\psi$ method[7].

Table 1: Experimental parameters employed in the X-Ray diffraction analysis

Wavelength Kα (Cr)	**0.2291 nm**	**Filter**	**Vanadium**
Exposure time (s)	20	Ø collimator (mm)	2
Tilt ψ (°)	9 points between -45/+45	Rotation angle, φ (°)	-45, 0 y +45
Background	Parabolic	Fit	Pseudo-Voigt
Measuring mode	χ-modified	Diffraction angle	156.0°
Miller indices (hkl)	(211)	Elastic constant, E/(1+ν) (GPa)	168.9 ± 2.8

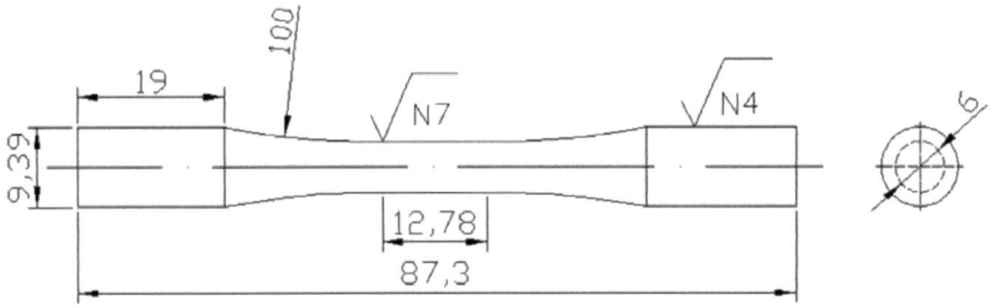

Fig. 1: Geometry and dimensions (mm) of the fatigue specimens

Diffraction data were determined in three different directions on the sample plane, -45, 0 and 45°, subsequently calculating the average result. Measurements were made on an X-Stress 3000 G3R device manufactured by Stresstech under the experimental conditions shown in Table 1. The slight stress relaxation produced by electropolishing layer removal was also taken into account and corrected in accordance with Sikarskie[8], who has developed a methodology based on the Moore and Evans procedure[9].

Roughness and residual stress profiles were obtained on flat specimens transversally cut from the aforementioned 16 mm bars.

Finally, fatigue tests were carried out on a four-point loading R. R. Moore rotating beam fatigue testing system (fully reversed loading, R = -1) in room temperature air at a nominal frequency of 95 Hz (5700 rpm). The geometry and dimensions of the fatigue specimens are shown in Fig. 1. The applied maximum alternative surface stress was 1100 MPa (55% of the tensile strength of the steel). This applied load was defined so that it was always above the fatigue limit and, in the case of the best treatments, not too far from it. The number of specimens tested under each condition varied between 4 and 6, with the fatigue results being expressed as average results, although individual values were also reported. Once the specimens had failed, fractographic analyses were carried out using a SEM microscope to detect the origin of the failure and the fracture pattern.

3.0. Results

3.1. Shot peening+vibratory finishing

Table 2 shows the evolution of the roughness parameters versus the time of application of the second, vibratory finishing treatment. Roughness decreased significantly in an approximately linear way with the time of application of the vibratory finishing treatment.

Table 2: Roughness evolution with time of second treatment

Shot Peening + Vibratory Finishing

	R_a (μm)	R_{max} (μm)
SP 19A	2.3 ± 0.2	15.8 ± 1.5
SP19A+ VF-8h	1.7 ± 0.1	11.5 ± 1.9
SP19A+ VF-16h	1.2 ± 0.1	9.4 ± 1.6
SP19A+ VF-24h	0.9 ± 0.1	8.0 ± 1.3

Shot Peening + Grit Blasting

	R_a (μm)	R_{max} (μm)
SP19A+GB-2s	2.0 ± 0.1	14.5 ± 0.7
SP19A+GB-10s	2.0 ± 0.15	15.0 ± 1.7
SP19A+GB-30s	1.7 ± 0.1	12.9 ± 1.1
SP19A+GB-60s	1.5 ± 0.1	12.2 ± 1.5
SP19A+GB-90s	1.5 ± 0.1	12.0 ± 1.3

Fig. 2 shows the residual stress profiles of the different double treatments, along with the profile corresponding to the single 19A shot peening treatment. In order to obtain a good estimate of the layer removed in the vibratory finishing treatments, all the profiles were superimposed on the stress decay region: the surface layer removed after 8 h of vibratory finising was 50 μm, 55 μm after 16 h and, finally, 60 μm after the 24 h treatment.

Fig. 3 shows the fatigue results obtained under a maximum alternative stress of 1100 MPa. The results obtained with non-treated specimens (NSP) and with specimens submited to the single 19A shot peening treatment are also included. Note that all the double treatments increased the fatigue life obtained following the single shot peening treatment. The progressive increase in fatigue life with the duration of the vibratory finishing treatment is also evident: the 24-h vibratory treatment increased the fatigue life obtained via the single shot peening treatment almost sevenfold.

3.2. Shot peening+grit blasting

Table 2 shows the evolution of the roughness parameters versus the time of application of the second, grit blasting treatment. In this case, roughness was barely affected by the grit blasting operation and grit blasting time did not significantly modify the roughness profiles.

Fig. 4 shows the residual stress profiles of the different double treatments, along with the profile corresponding to the single 19A shot peening. As in Fig. 2, the profiles were superimposed on the stress decay region. It can be seen that the surface layer removed after 6, 10, 30 and 60 s of grit blasting treatment were 30, 50, 70 and 120 μm, respectively.

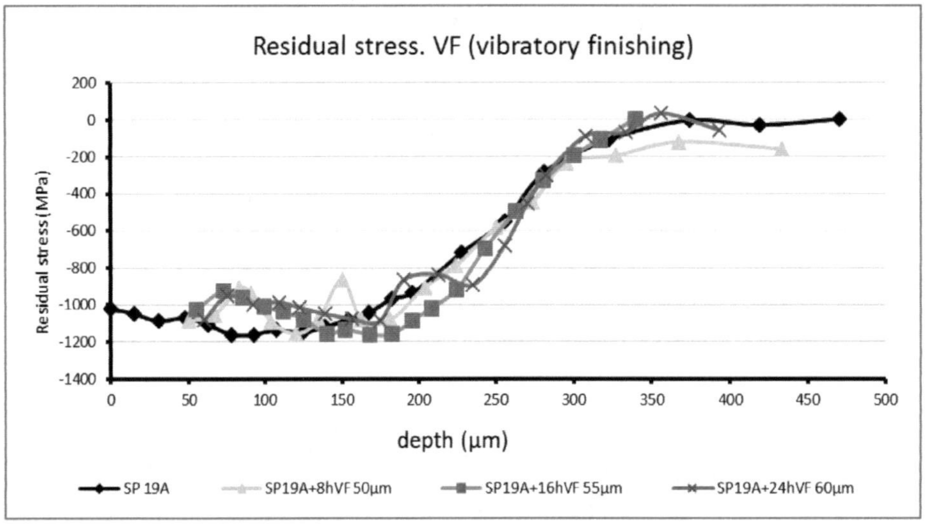

Fig. 2: Residual stress profiles of SP19A and SP19A plus the different second, vibratory finishing treatments. Uncertainty of each measurement: 2% or ± 50 MPa

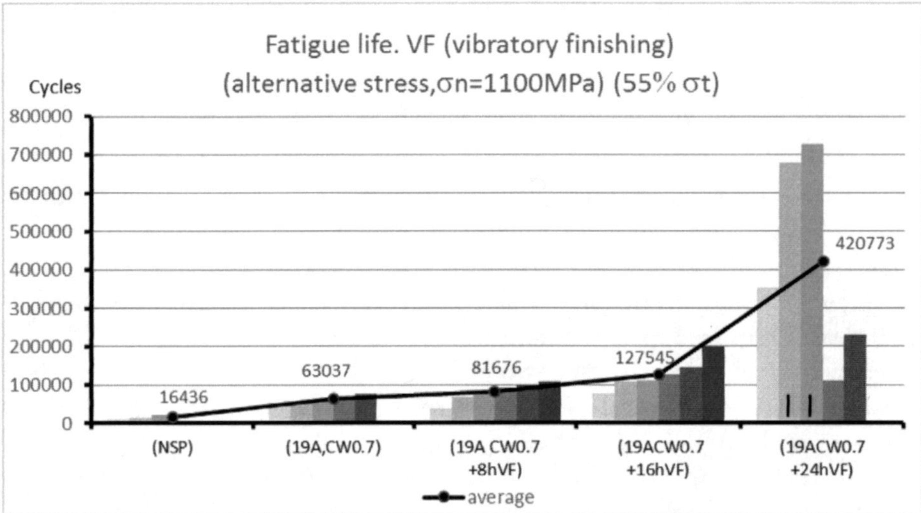

Fig. 3: Fatigue life under an alternative stress of 1100 MPa (55% ultimate tensile strength) of double treatments with a second, vibratory finishing treatment. "I" denotes internal fatigue crack initiation. Fatigue crack initiated on the surface in all the other specimens

Fig. 5 shows the fatigue results obtained under a maximum alternative stress of 1100 MPa. Note that all the double treatments increased the fatigue life obtained after the single shot peening treatment. The progressive increase in fatigue life with the duration of the grit blasting treatment is also clear up until the 60 s treatment. However, fatigue life decreases with the 90 s treatment because an overly deep surface layer was removed and the region under compressive stresses was excessively reduced. In this case, the 60 s grit blasting double treatment increased the fatigue life obtained via the single shot peening treatment twelvefold.

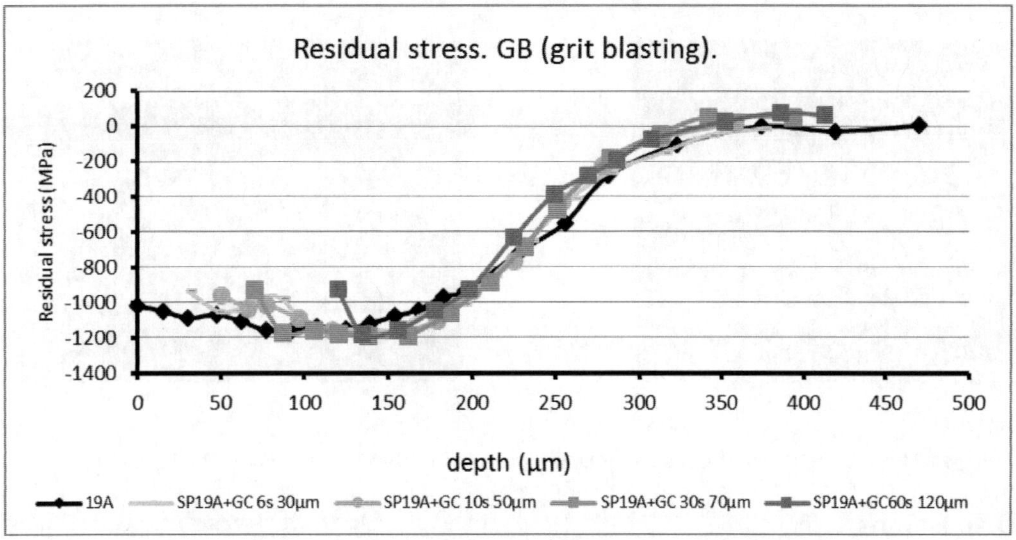

Fig. 4: Residual stress profiles of SP19A and SP19A plus the different second, grit blasting treatments. Uncertainty of each measurement: 2% or ± 50 MPa

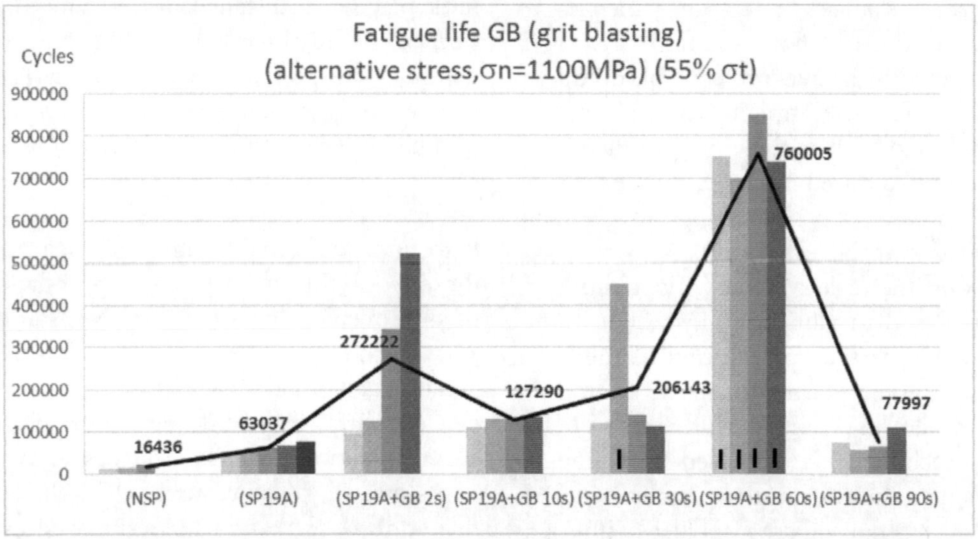

Fig. 5: Fatigue life under an alternative stress of 1100 MPa (55% ultimate tensile strength) of double treatments with a second, grit blasting treatment. "I" denotes internal fatigue crack initiation. Fatigue crack initiated on the surface in all the other specimens

This treatment removed a layer of about 60 µm from the fatigue specimen for the aforementioned grit blasting time. The layer removed from the fatigue specimens was half that removed from flat samples, because half of the specimen surface lies outside the grit blasting projection angle.

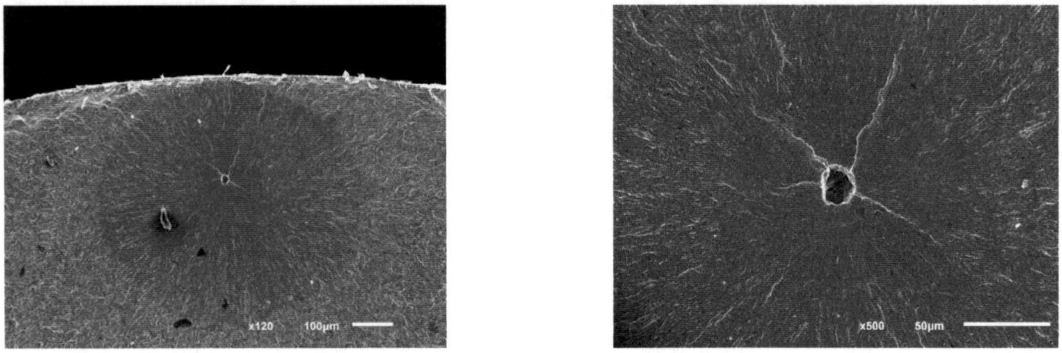

Fig. 6: Fracture surface: internal crack initiation at an alumina inclusion (SP19A+GB60s)

4.0. Discussions

Double surface treatments, consisting in first applying a high intensity peening treatment, the major effect of which is to produce a deep region submitted to high magnitude compressive stresses, followed by a second surface treatment to reduce the roughness induced in the first treatment and mitigate the induced damage, proved to be highly suitable for increasing the fatigue performance of the tested steel.

On the other hand, grit blasting is much better than vibratory finishing in terms of efficiency, as the optimization of the latter technique requires a treatment of around 24 hours, while the best performance of the former technique is achieved in only 60 seconds. A second treatment by means of grit blasting using fine alumina powder (with a grit size below 0.1 mm) applied for 60 s onto the fatigue specimens removes an average surface layer of about 60 μm, removing the largest defects induced in the previous shot peening process, while leaving a compressive residual stress depth in the range of 250-300 μm.

It should be noted that, as shown in Figs. 3 and 5, the initiation of fatigue cracks was internal in the case of the specimens that had the longest fatigue lives, although the maximum stress in our fatigue experiments (bend testing) always acted on the specimen surface. One of these fracture surfaces, observed under the scanning electron microscope, can be seen in Fig. 6. The location of fatigue damage initiation, which took place at a distance of 250 μm from the surface, is clearly visible in this figure. This depth corresponds to a point where the residual compressive stress had been reduced to almost zero in this specimen following the applied double treatment. Fatigue damage initiation took place due to the presence of a hard, non-deformable alumina inclusion. In these cases, the stress concentration factor due to shot peening surface defects was minimized and the compressive stress belt was able to avoid crack initiation on the surface region, consequently triggering an internal fatigue crack initiation process. In these cases, the stress intensity factor provided by the sum of the applied (σ_a) and the residual stress (σ_r) acting on the largest surface defect was lower than the subsurface stress intensity factor (lower applied bending stress, but also much lower compressive residual stress) acting on the largest inclusions present in the steel (see Fig. 7). The so-called internal fatigue damage mechanism is non-dependent on surface characteristics, depending only on the steel microstructure, including the existence of microstructural stress concentrators such as alumina inclusions.

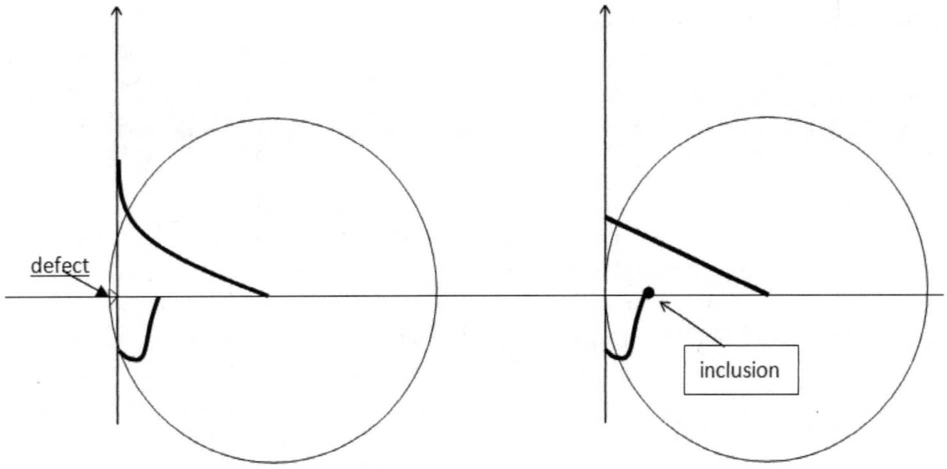

Fig. 7: Stresses under bending and residual stresses on shot peened fatigue specimens. **(a)** Specimen with surface defects. **(b)** Specimen after removal of surface defects (double surface treatment): fatigue crack initiation at a subsurface inclusion

All the fatigue specimens tested after the double treatment with second, grit blasting treatment for 60 s lasted more than 700,000 cycles, with initiation of the fatigue crack taking place internally (see Fig. 5). In the case of the double treatments with a second, vibratory finishing treatment maintained over 24 hours, however, only two out of the five specimens showed internal fatigue initiation (both also lasted about 700,000 cycles), while the fatigue initiation location of the other three specimens (with a significant lower fatigue life) took place on the specimen surface. In this case, the second treatment was not able to remove all the shot peening defects completely, even though the lowest roughness parameters were obtained using this technique. It would appear that vibratory finishing should be prolonged for a longer time in order to remove the shot peening defects completely. However, a treatment of longer than 24 hours (the longest employed in this study) is not convenient from either a practical or an economic point of view.

5.0. Conclusions

- Double surface treatments, consisting in first applying a high intensity peening treatment followed by a second surface treatment, vibratory finishing or grit blasting, proved to be highly suitable for increasing the fatigue performance of structural steels.
- In this respect, grit blasting was found to be more efficient than vibratory finishing, as the optimization of the latter technique requires a treatment lasting around 24 hours, while the best performance using the former technique is achieved in only 60 seconds. Moreover, the increase in fatigue life of the steel is also higher in the former case.
- It was shown that the initiation of fatigue cracks were internal in the case of the double treated specimens that had the longest fatigue lives (optimal treatments). The location of fatigue damage initiation took place at a point where the residual compressive stress was reduced to almost zero. Furthermore, fatigue damage initiation took place due to the stress concentration provided by the presence of hard, non-deformable alumina inclusions.
- The real internal fatigue life of the steel was attained via the application of an optimal surface double treatment. This is

non-dependent on surface characteristics, depending only on the microstructure of the steel, including the existance of microstructural stress concentrators such as alumina inclusions.

6.0. Acknowledgments

The authors are grateful for the financial support for this study provided by the European Union (ERDF funds) and Principado de Asturias, through the Plan de Ciencia, Tecnología e Innovacion, 2013-2017 (FC-15-GRUPIN14-001 Project).

7.0. References

1. Weich, I.; Ummenhoffer, T.; Mater. Manuf. Process., 2011, vol. 26, p. 288.
2. Fernandez Pariente, I.; Guagliano, M.; Mater. Manuf. Proces., vol. 24, p. 1436.
3. Tekeli, S.; Mater. Lett., 2002, vol. 57, p. 604.
4. Gao, Y.K.; Surf. Eng., 2008, vol. 24(5), p. 322.
5. Vielma, A.T.; Llaneza, V; Belzunce, F.J.; Surf. Coat. Technol., 2014, vol. 249, p. 75.
6. Kawagoishi, N.; Nagano, T.; Moriyama, M.; Kondo, E.; Mater. Manuf. Process., 2009, vol. 24, p. 1431.
7. Noyan, I.C.; Cohen, J.B.; Residual stress: Measurement by diffraction and interpretation, MRE, Springer-Verlag, New York, 1987.
8. Sikarskie, D.L.; On a series form of correction to stresses measured using X-ray diffraction, AIME Transactions, 1967, p. 239.
9. Moore, M.G.; Evans, W.P.; Mathematical correction for stress in removed layers in X-ray diffraction residual stress analysis, SAE Transactions, 1958, p. 66.

Hydrophobic Anti-Ice Coatings to Prevent Frost Spreading

S. Holberg*, R. Losada, J.W. Kristoffersen,
Danish Technological Institute,
Kongsvang Allé 29, 8000 Aarhus C, Denmark.

K. Kamuk, F. Nielsen,
Exhausto A/S, Odensevej 76, 5550 Langeskov, Denmark.

N. Arconada, M. Hernaiz, R. Ortiz, E. Rodriguez
IK4-Tekniker, Iñaki Goenaga 5, 20600 Éibar, Spain.
*Email: SHG@teknologisk.dk

Abstract

Heat exchangers for refrigeration, heat pumps and heat recovery ventilation (HRV) require periodic defrosting which consumes energy. Application of a smooth, hydrophobic, organic-inorganic hybrid coating as wet paint, which cured provides low contact angle hysteresis (advancing/receding water contact angles 104°/95°), extended the time interval between defrosting cycles in a HRV experiment by a factor of 2.3. Freezing (ice nucleation) is not inhibited, but the spreading of frost, which originates from ice-nucleation sites, was inhibited. Herein we discuss suitable analytics, applied coating systems and challenges for further improvement by structured, superhydrophobic surfaces.

1.0. Introduction

The present work addresses icing problems of heat recovery ventilation (HRV) in modern, energy efficient buildings. Aluminium counter-flow plate heat exchangers warm the incoming cold air with the outgoing warm air. At outside temperatures below about -3 °C, the more humid outgoing air is cooled to below 0 °C. Ice accumulates over time, blocking the flow. Therefore, the unit requires periodic defrosting by heating, which consumes energy. Similar problems occur for refrigeration or heat pumps. While frost build-up may not be completely avoided, longer cycles between defrosting cycles would save energy.

'Anti-ice' is an umbrella term for several different physical effects such as reduced ice adhesion, inhibition of ice nucleation (the freezing of liquid water) and/or deceleration of the spreading of frost from a frozen spot on a cold surface. Ice nucleation occurs randomly and with a distinct probability related to temperature and surface area, but not at a distinct temperature. The ice nucleation probability increases with increasing surface area and decreasing temperature. Ice nucleation is inhibited in pure bulk water, while solid surfaces promote freezing[1]. For heat exchangers, usually freezing inhibition is not considered. However, previous investigations at temperatures of -4 to -7 °C revealed a comparably low ice nucleation probability for both a hydrophobic coating (herein called C1a) and bare Al. Wet, 100 cm^2 large surfaces stayed ice-free in some test runs for several hours[2].

Once ice nucleation occurs on a cold surface that is covered by a continuous water film, ice forms instantly on the complete surface. On hydrophobic surfaces, water forms discrete drops that are not connected with each other. Ice nucleation leads to the freezing of one single drop. However, at the same temperature, a frozen drop has a lower vapour pressure than a liquid drop. This leads to the frozen drop growing faster due to deposition of humidity, while a neighbouring liquid drop might shrink. The frozen drop eventually touches the neighbouring liquid drop and induces freezing. This mechanism still leads to frost spreading, but at a low rate[3]. The ability of a shrinking liquid drop to contract and thus move away from a neighbouring frozen drop is determined by the receding contact angle. Kim et al. related spreading to the critical diameter D_c, which characterises the volume that a condensing drop must reach before it starts to slide down a surface. D_c can be calculated from advancing and receding contact angles according to the formula provided by Fig. 1. Low contact angle hysteresis (CAH) and high absolute contact angles, which in combination require a high receding contact angle, lead to a low D_c and thus to slow frost spreading[4].

Herein we report various attempts to prepare suitable surfaces and evaluate an accordingly coated HRV heat exchanger. Several attempts involve organic-inorganic hybrid coatings prepared by sol-gel processing. Ethoxy-or methoxy-functional silanes are hydrolysed and subsequently condensed to form a siloxane network. To obtain a balance between the rigid siloxane network providing chemical stability and the flexibility or organic coatings, silanes may comprise larger organic substructures. For example, the coatings herein called C1a/C1b are prepared from a mixture of methyltriethoxysilane and the adduct of (3-aminopropyl)methyldiethoxysilane and 4,4'-methylenebis(cyclohexylisocyanate) as shown by Fig. 2 [5].

To obtain a hydrophobic surface, an approach that has been well established for the preparation of easy-to clean coatings is applied.[6] Either polydimethylsiloxane (PDMS, silicone) or perfluoroalkyl components were added to the coating resins, both comprising further functional groups for chemical cross-linking.

$$D_c = \sqrt{\frac{24\,(\sin\theta_{adv})^3\,\gamma(\cos\theta_{rec} - \cos\theta_{adv})}{\pi\rho(1-\cos\theta_{adv})^2\,(2+\cos\theta_{adv})\,g\sin\alpha}}$$

α: surface tilt angle (vertical = 90°)
ρ: density of the liquid
g: gravity constant
γ: surface tension of the liquid
θ_{adv}, θ_{rec}: advancing/receding contact angles

Fig. 1: Formula for the critical diameter D_c at which a drop starts to slide down according to Kim et al.[4]

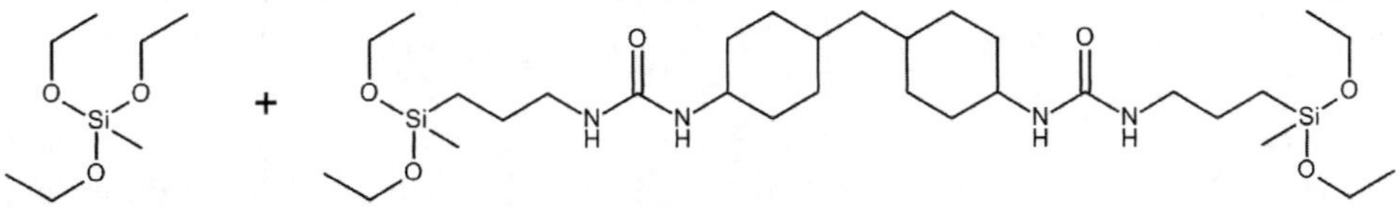

Fig. 2: Silanes applied to prepare coatings C1a and C1b.

When choosing appropriate components, a surface-tension-driven separation process leads to orientation of the hydrophobic moieties towards the surface providing hydrophobicity and repellence as illustrated by Fig. 3. Basic principles of the separation process have been described by Madjumdar & Webster[7] and Esteves et al.,[8] as well as the effect of different PDMS chain lengths on the coating herein called C1 by Holberg & Bischoff[5].

2.0. Methods

Materials: Plates: Al8006 (0.2 mm) or Al3003H14 (0.7 mm), both bare mill finish. Coatings: C1a/C1b are previously published organic-inorganic hybrid coatings,[9] C2a/C2b are further developments of C1a/C1b. The "a" and "b"-versions are identical except that "a"-versions comprise a PDMS (polydimethylsiloxane) additive that cross-links with the coating matrix. C3a is a sol-gel coating based on TEOS (tetraethyl orthosilicate) and PFOS ((1H, 1H, 2H,2H-perfluorooctyl)triethoxysilane), C3S is the same coating as C3a, but the Al surface was structured with a laser (line grid, peak-valley ~40 µm, peak-to peak ~70 µm) prior to application of the coating. PDMS is bulk silicone rubber (=cross-linked PDMS).

All coatings were applied by spraying, except C3a, C3S and the application onto the HRV heat exchanger which were by dipping. Curing: C1a/b: 1 h at 200 °C, C2a/b on test samples: 0.5 h at 60 °C followed by 1 h at 140 °C, C2a on the heat exchanger: 13 h at 60 °C, C3a/S: 150 °C, PDMS: 24 h at 20 °C. Thickness: C1a/b, C2a/b: 4-8 µm, C3a/S below 1 µm, PDMS: 15-30 µm. Polymer films: FEP: Fluorinated ethylene propylene, 25 µm foil (Bohlender), PP: Polypropylene film, 50 µm (VWR). M1 is a monolayer prepared by immersing Al for 10 min in ~85 °C warm water, followed by drying, immersion in a 1% solution of PFOS in Ethanol for 3 d at ~20 °C, rinsing and tempering for 1 h at 120 °C. M2 is a perfluoroalkylsilane monolayer applied by a PVD process.

Contact angles: Determined with a DSA 10 (Krüss), T~20 °C, five measurements per data point. Video sequences were not recorded and therefore the reported advancing and receding contact angles may be slightly lower and higher, respectively, compared to the actual values.

Frost spreading rate: The test chamber was previously described[2]. Sample plates (101 x 152 mm) were mounted vertically inside a test box and cooled from their backside.

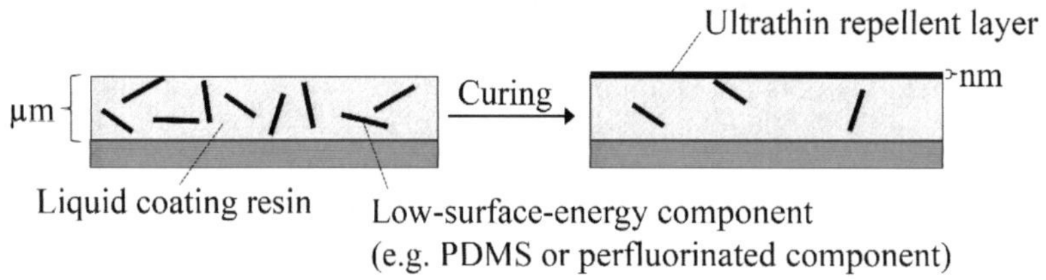

Fig. 3: Surface-tension-driven orientation process to obtain hydrophobic surfaces by hydrophobic additives

While the sample plate edges were hold above 0 °C, the area between lines 12 mm from the edges (~100 cm²) was cooled to -4.0±0.5 °C. Temperature accuracy was determined by thawing tests. The box atmosphere was +12 °C/90% relative humidity, and in the case of forced air flow: 60% humidity. Water condensed continuously on the plates. When water started to run down in drops, the tests were commenced by placing ice crystals in the centre of the plates. Approximately every 10 min within the first 40 min, the frozen area was evaluated. A growth rate of the radius of a circle enclosing an identical area as the actual frozen area was calculated. The growth rate of a single test run was found to be constant (±20%) during the first 40 min.

3.0. Results and Discussion

Hydrophobic surfaces were prepared by various techniques. First, we discuss smooth surfaces, which herein refer to all surfaces without an intentionally manufactured micro- or nanostructure (all except C3S and M1). According to the wetting characteristics provided in table 1, both coatings with a hydrophobic surface additive (C1a, C2a, C3a), monolayers (M2) and hydrophobic bulk polymers (FEP, PDMS) can provide a low D_c due to high contact angles and/or low CAH and are thus promising candidates to decelerate frost spreading. The hydrophobicity of all mentioned surfaces is provided either by PDMS or by perfluoroalkyl components. The investigated PP-sample provided less hydrophobicity. According to Andrade et. al., CAH occurs for multiple reasons, including surface inhomogeneity and roughness, the liquid leading to reorientation or deformation of the surface or the liquid being absorbed.[10] For the investigated smooth surfaces, C1a and C2a provided the lowest CAH. We speculate that this is due to a combination of low deformation (compared to the softer bulk polymers) and a rather homogeneous, defect-poor surface (compared to the monolayer).

Immersion in water for three days, simulating contact with humidity, led to a slight decrease of hydrophobicity for all surfaces. The interfacial tension between air and a solid surface favours the orientation of structural units that provide low surface tension, such as CH_3 or CF_3 groups, towards the solid-air interface. When the solid is in contact with water, this driving force is gone and reorientation may occur within the limits set by rigidity of the coating network.[11] However, the effect of the PDMS additive was maintained when comparing C1a/C2a with C1b/C2b after immersion. Static contact angles (not shown) were close to the advancing (within +2/-5°).

For selected surfaces, we determined frost-spreading rates. The results obtained by placing ice on -4 °C wet surfaces are provided by Table 1 and illustrated by Fig. 4. We took a significantly larger surface area into account than any previous investigation [3,12].

Table 1: Dynamic water contact angles and frost spreading rates. First values are initial, second values in brackets are after immersion in water for three days at ~20 °C and wiping dry. D_c is calculated for a vertical surface according to literature [4]. The frost-spreading rate of C1a is based on four, the other rates on one experiment

Coating	Water contact angles adv.	rec.	CAH	D_c	Frost spreading rate
bare Al	96.3±2.2°	50.3±5.9°	46°	4.2 mm	3000 μms^{-1}
C1a	106.2±0.8° (105.2±0.8°)	95.8±1.5° (89.8±1.5°)	10° (16°)	1.8 mm (2.3 mm)	2±1 μms^{-1}
C1b	90.0±1.0° (81.4±1.1°)	70.4±3.8° (52.4±0.5°)	20° (29°)	3.1 mm (4.0 mm)	21 μms^{-1}
C2a	104.2±1.3° (102.8±0.4°)	94.6±2.9° (86.8±1.1°)	10° (16°)	1.8 mm (2.3 mm)	2 μms^{-1}
C2b	80.2±2.6° (76.4±2.1°)	54.0±2.3° (34.2±0.9°)	26° (52°)	3.9 mm (5.2 mm)	
C3a	111.8±1.6° (106.0±1.4°)	93.8±4.8° (83.2±2.5°)	18° (23°)	2.1 mm (2.7 mm)	3 μms^{-1}
C3S	150.6±2.6°	138.0±2.7°	13°	0.5 mm	
M1	114.4±1.1° (112°)	103.8±3.3° (91°)	11° (21°)	1.5 mm (2.3 mm)	
M2	122° (117°)	88° (79°)	34° (38°)	2.4 mm (2.8 mm)	
FEP	113.5±2.4° (113.5±3.1°)	95.0±3.6° (87.5±5.4°)	19° (26°)	2.1 mm (2.5 mm)	
PDMS	112.0±0.8° (111.4±1.2°)	89.7±0.5° (79.1±3.1°)	22° (32°)	2.4 mm (2.9 mm)	
PP	96.2±3.1° (93.6±1.1°)	59.0±3.3° (45.2±1.1°)	37° (48°)	3.9 mm (4.4 mm)	

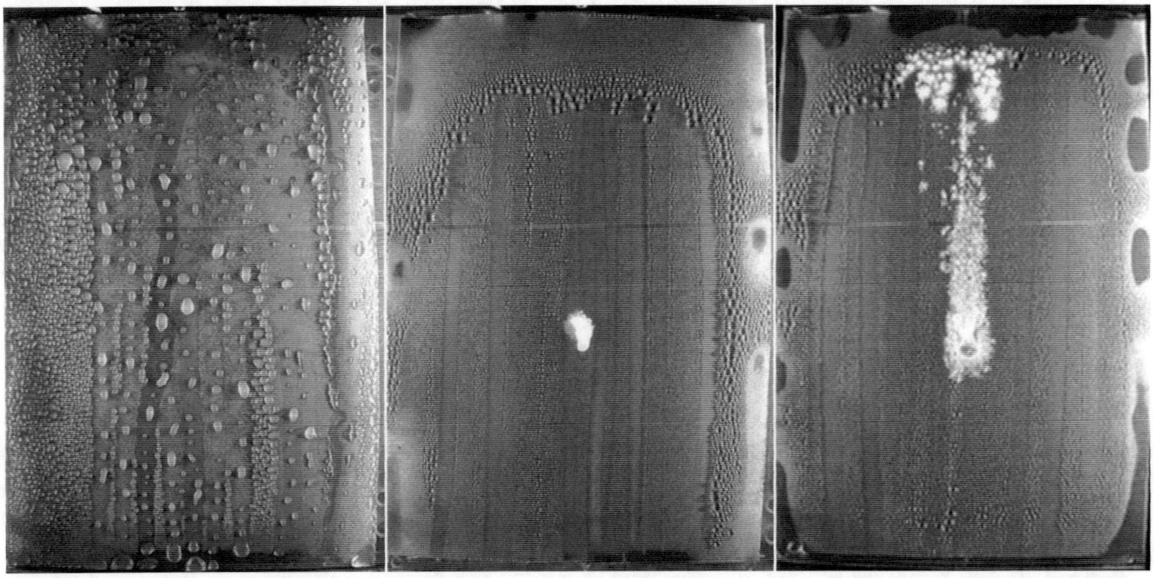

Fig. 4: Frost spreading on -4 °C wet surfaces. Freezing was initiated in the centre of each of the three plates at t = 0. Left: Bare Al after 1 min, opaque drops/areas are frozen. Middle and right: C1a after 20 min with natural convection (middle) or with an upward air flow of 1 ms^{-1} (right). Frost (white areas) is apparent.

Our data supports the previously proposed connection of D_c and the frost spreading rate with the coatings C1a, C2a and C3a with their low D_c leading to spreading rates in the range of several μms^{-1}. For the first time, a surface with moderate hydrophobicity was tested (C1b), showing a moderate spreading rate. On bare Al, frost spreading was exceptionally fast.

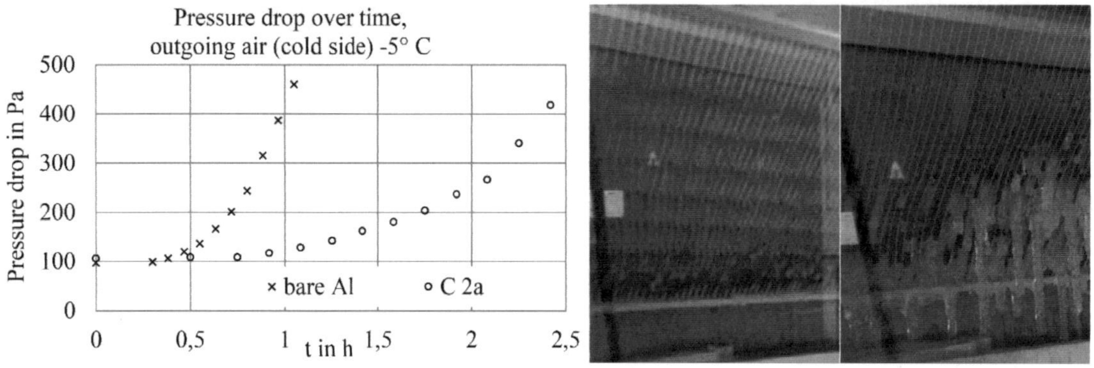

Fig. 5: HRV test according to EN 308. The outgoing, cooled air leaves the heat exchanger at -5 °C. Left: Pressure drop over time. Middle/Right: Outlet of an uncoated (middle) and a coated (right) heat exchanger.

Despite a receding contact angle above 0°, we expect a continuous water film as previously proposed[13].

When repeating the experiment with an airflow of 1 ms^{-1}, which is realistic for HRV applications, frost spreads fast in the direction of the flow (see Fig. 4, right). Even though condensation happens all over the complete surfaces, humidity, and thus also condensation rate, is highest where the ventilated air meets the cold plate and decreases in the airflow direction. The present frost-spreading model suggests growth into the direction of the humidity, which would be in opposite direction of the flow, and does not explain the observed phenomenon. At present, the authors propose that the transport of loose frost by the airflow is the only plausible explanation.

In a HRV experiment, according to EN 308, two heat exchangers were tested, one with a bare Al surface, the other coated with C2a. Within the measuring accuracy of the measurement, the coating did not decrease the heat transfer through the Al plates. During the experiment, the flow was kept constant by adjusting the fan power. Frost leads to an increasing pressure drop. Once the pressure reached 400 Pa, defrosting would usually start. As shown by Fig. 5, despite the consequences of the airflow discussed above, C2a increased the time between defrosting by a factor of 2.3. While frost builds up inside the channels of the uncoated heat exchanger, many channels of the coated heat exchanger are ice-free. Liquid water leaves the channels of the coated heat exchanger, eventually forming icicles (see Fig. 5, right). The icicles do not block the flow to the same extend as the frost inside the channels of the heat exchanger.

Structured, superhydrophobic surfaces are promising candidates to further reduce D_c and frost spreading rate compared to smooth hydrophobic surfaces due to low CAH and high contact angles. However, these structured surfaces behave differently under condensation than when applying mm-sized water drops. Standard contact angle measurements are thus not sufficient to predict the surfaces' performance. Condensation leads to the nucleation of small water droplets inside surface structures. This leads to hydrophilicity or at least to reduced hydrophobicity. In the latter case, growing drops have partly full surface contact (Wenzel state) and are partly in an air-pocket state (Cassie Baxter) [13–15]. However, single superhydrophobic surfaces have been reported to be effective even under condensation[16,17].

Fig. 6: Samples at +1 °C during water condensation. Left to right: Bare Al, C1a, C3S and M1. The pipette tip is in each case guiding a drop from left to right to simulate a dynamic contact angle measurement

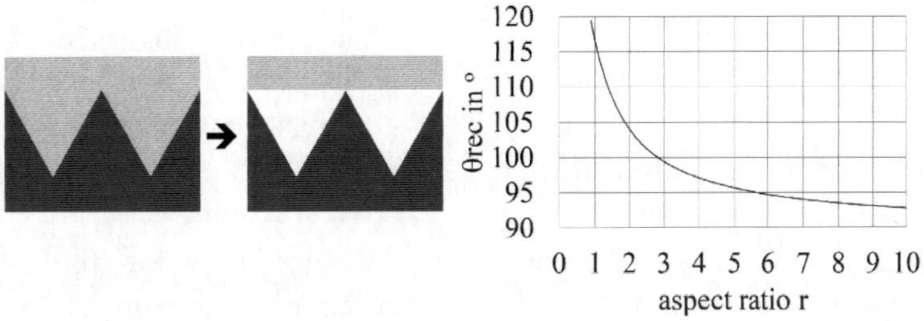

Fig. 7: Required surface chemistry (θ_{rec} derived from a smooth surface) in relation to the height/width aspect ratio of a square-based pyramidal surface for a Wenzel-to-Cassie-Baxter transformation

We prepared two structured, potentially superhydrophobic surfaces by either combining a microstructure prepared by laser with a hydrophobic coating (C3S) or by applying a silane monolayer to Al etched with hot water (M1). We did not further characterize the structure of M1; according to literature, hot water etching leads to a boehmite nanostructure [18]. Both structured and smooth surfaces listed in Table 1 were exposed to condensation conditions by keeping the samples at +1 °C in a horizontal position. After water had visibly condensed on the surfaces, a drop was guided with a pipette tip over the wet surfaces to simulate dynamic contact angles. Within the limits of such observation by bare eye, contact angles on dry surfaces (cf. Table 1) and under condensation conditions were identical for all smooth surfaces except bare Al. Bare Al and the structured surfaces C3S and M1 become more hydrophilic under condensation (see Fig. 6 in comparison to the values of Table 1). Under condensation, C3S and M1 provide receding contact angles of roughly about 90°, while the advancing angles are still above 110° (M1) or 120° (C3S). Both CS3 and M1 show, even under condensation, characteristic reflections at the bottom of the drops that, at least partly, indicate an air-pocket state. However, both structured surfaces are, with respect to low CAH, outperformed by smooth surfaces like C1a.

An ideal solution for applications involving condensation could be surfaces enabling a complete Wenzel-to-Cassie-Baxter transformation. We performed a theoretical calculation to relate surface chemistry with surface topology required for such transformation. The surface chemistry is represented by the contact angle of an otherwise identical, but smooth surface. As droplets initially are in contact and shall withdraw from a structure, the receding contact angle is applied. As topology, adjacent ideal square-based pyramids were selected. The topology is characterized by an aspect ratio r calculated by dividing the height of the pyramids (base to apex) by the width (=length of the base edge). We disregarded gravity. The approach 'Wenzel' = 'CB' for the surface energy of the structure is accordingly: E(lateral surface coating-water) = E(lateral sur face coating-air) + E(base surface air-water).

Fig. 7 (left) shows a 2d-scheme of Wenzel and Cassie-Baxter state. Applying the chosen geometry and the Young's equation for the contact angle, transformation leads to an expression to calculate the threshold (=minimum) receding contact angle of an otherwise identical, but smooth surface in dependence of the aspect ratio r: $\theta_{rec\ threshold} = \arccos(-(4r^2 + 1)^{-0.5})$. Fig. 7 (right) shows a graph of this equation. A high aspect ratio leads to lower $\theta_{rec\ threshold}$. However, even an infinite aspect ratio requires a surface chemistry that on a smooth surface leads to a receding water contact angle of at least 90°, not by average, but continuously. Receding contact angles up to 116° have been reported for smooth surfaces[19], but the present coatings do not fulfil this requirement after three days water immersion.

4.0. Conclusion

- Smooth hydrophobic coatings are suitable anti-ice surfaces. The coatings do not depress the ice nucleation, but decelerate the spreading of frost and improve drainage. Even though an airflow compromises the anti-ice properties to a certain extent, the effect was sufficient to extend the time between defrosting cycles in heat recovery ventilation (HRV) by a factor of 2.3, which would lead to energy savings in a real application. Additional potential applications, likewise utilising periodic defrosting, are heat pumps, refrigeration and wind turbines.
- With respect to technical applications, future research should address the long-term durability in the range of years in contact with humidity. In the present investigation, water immersion for three days lead to a significant decrease of the contact angles of all investigated surfaces.
- Low CAH and high absolute water contact angles favour slow frost spreading. This previously proposed relation was verified applying improved test methods. Among the investigated surfaces, organic-inorganic hybrid coatings comprising a PDMS additive to achieve surface hydrophobicity lead to the most promising wetting characteristics with respect to decelerate frost spreading. However, the contact angle data suggests that monolayers, hydrophobic bulk polymers and further coatings based on perfluoroalkylsilanes should as well have an ability to decelerate frost spreading.
- The attempts of the present work to prepare structured, superhydrophobic surfaces did not lead to an improvement compared to the best smooth surfaces. However, the structured surfaces partly maintained an air-pocket (Cassie-Baxter) state during condensation. A complete Wenzel-to-Cassie-Baxter transformation for growing water drops seems quite challenging. Calculations show that not only are surface topologies with high slopes required, but also a surface chemistry that on a smooth surface already leads to a receding water contact angle above 90°.

5.0. Acknowledgement

The authors thank the European Union for financial support through the FP 7 project ENE-HVAC, K. B. Haugshøj for preparing the PVD samples, N. Tauber for helping to design the ice test chamber and F. H. Blaikie for revising the manuscript.

6.0. References

1. Vali, G.; 'Ice nucleation - Theory,' Paper presented at the NCAR/ASP 1999

Summer Colloquium, Boulder, CO, USA, 1999.
2. Holberg, S.; Losada, R.; Anti-ice: 'What's possible? - What's similar to other functional coatings?' Paper presented at the European Coatings Congress 2015, Nürnberg, Germany, 2015 (April).
3. Zhang, Q.; He, M.; Zeng, X.; Li, K.; Cui, D.; Chen, J.; Wang, J.; Song, Y.; Jiang, L.; Soft Matter, 2012, vol. 8, p. 8252.
4. Kim, P.; Wong, T-S.; Alvarenga, J.; Kreder, M.J.; Adorno-Martinez, W.E.; Aizenberg, J.; ACSNano, 2012, vol. 6, p. 6569.
5. Holberg, S.; Bischoff, C.; Progress in Organic Coatings, 2014, vol. 77, p. 1591.
6. Sepeur, S.; Laryea, N.; Goedicke, S.; Groß, F.; Nanotechnology - Technical Basics and Applications, Vincentz Network, Hannover, Germany, 2008, p. 39ff.
7. Majumdar, P.; Webster, D.C.; Polymer, 2007, vol. 48, p. 7499.
8. Esteves, A.C.C.; Lyakhova, K.; Van der Ven, L.G.J.; Van Benthem, R.A.T.M.; De With, G.;: Macromol., 2013, vol. 46, p. 1993.
9. Bischoff, C.; Holberg, S.; 'Repellent coating composition and coating, process for making and uses thereof,' examples 1 and 2, Patent WO 2012/083970 A1, 2012.
10. Andrade, J.D.; Smith, L.M.; Gregonis, D.E.; Surface and Interfacial Aspects of Biomedical Polymers, J. D. Andrade, ed., Plenum Press, New York, NY, USA, 1985, p. 249.
11. Li, Z.; Han, W.; Kozodaev, D.; Brokken-Zijp, J.C.M.; De With, G.; Thüne, P.; Polym., 2006, 47, p. 1150.
12. Chen, X.M.; Ma, R.; Zhou, H.; Zhou, X.; Che, L.; Yao, S.; Wang, Z.; Scientific Reports, 2013, vol. 3, p. 2515.
13. Dooley, J.B.; 'Determination and characterization of ice propagation - Mechanisms on surfaces undergoing dropwise condensation,' PhD thesis, Texas A&M University, College Station, TX, USA, 2010.
14. Wier, K.A.; McCarthy, T.J.; Langmuir, 2006, vol. 22, p. 2433.
15. Narhe, R.D.; Beysens D.A.; Langmuir, 2007, vol. 23, p. 6486.
16. Boryeko, J.B.; Chen, C-H.; Physical Rev. Lett., 2009, vol. 103, p. 184501.
17. He, M.; Wang, J.; Li, H.; Jin, X.; Wang, J.; Liu, B.; Song, Y.; Soft Matter, 2010, vol. 6, p. 2396.
18. Kim, P.; Kreder, M.J.; Alvarenga, J.; Aizenberg, J.; Nano Lett., 2013, vol. 13, p. 1793 (supporting information).
19. Nishino, T.; Meguro, M.; Nakamae, K.; Matsushita, M.; Ueda, Y.; Langmuir, 1999, vo. 15, p. 4321.

The Impact of Zeolite Application Technique on the Surface Morphologies of Cellulose Fibres

D. Fakin*,
University of Maribor,
Faculty of Mechanical Engineering,
Institute for Engineering Materials and Design,
Smetanova 17, 2000 Maribor, Slovenia.

A. Horvat,
Silkem, d.o.o., Tovarniška cesta 10,
2325 Kidričevo, Slovenia.

A. Ristić,
[c]National Institute of Chemistry,
Hajdrihova 17, 1000 Ljubljana, Slovenia.

S. Hribernik, A. Ojstršek*
University of Maribor,
Faculty of Mechanical Engineering,
Institute for Engineering Materials and Design,
Smetanova 17, 2000 Maribor, Slovenia.
*Email: darinka.fakin@um.si, alenka.ojstrsek@um.si

Abstract

The purpose of presented paper was to evaluate the impact of two different industrially-viable procedures as well as numerous operational parameters and baths' compositions on the surface morphologies of cellulose fibres and consecutively on the UV-rays blocking abilities of zeolite-modified fabrics in comparison with the untreated ones. The obtained SEM, FTIR and X-ray diffraction analyses results of zeolite particles revealed the presence of three different types of zeolites within the examined sample, i.e. the largest amount being cubic crystals of zeolite A, followed by the octahedral crystals of zeolite X, and the pentasil building blocks of zeolite ZSM-5. In addition, the most sufficient application of those zeolites mixture was achieved with the pad-dry-thermofix process using 30 g L^{-1} of zeolites with or without the presence of a crosslinking agent, which was confirmed by SEM and FTIR analyses indicating fabrics' good to very good UV protective properties. On the other hand, the exhaustion application technique resulted in a lower amount of particles onto the surface of cotton fabric, depending on the zeolites' concentrations and the chemical agents used.

1.0. Introduction

Increasing requests for functional, comfortable and safe fibre-forming materials, as well as major global competition on this field, prescribe the intensive development of new technologies for the processing of high added-value textiles and the developments of innovative but low-cost and environmentally-friendly procedures for their finishing. Recently, there has been an augmented demand for textiles and apparel from materials of natural origin that provide greater comfort, and remain fresh and odour free after prolonged usage[1]. Moreover, the use of natural cellulose-based materials has recently become of great interest for numerous new fields of application (protective clothing, clothing for medical and sport activities, interior decorative coverings, technical textiles, etc.), due to the fact that they are readily available in large quantities and easily produced, and because their molecular structures offer excellent potential as a matrix for the designing of bioactive, biocompatible, and intelligent materials[2,3]. In spite of various excellent properties, i.e. high flexibility, high mechanical stability, low density, etc., the usage of cellulosic textiles for special applications is limited by certain disadvantages, especially a high level of combustibility, lack of antimicrobial activity, and in some cases excessive water absorption ability. Thus, the modification of cellulose fibres is required. Numerous techniques are presently being employed for changing the materials' surface morphologies, and consecutively their properties, i.e. dip-coating by the dispersion of well-defined nano-oxides[4–6], chemical grafting of particles[7], functionalization by inorganic sol-gel coating[8], and layer by layer deposition method[9,10]. Therefore, it is significantly important to establish an application technique that can be readily implemented by textile manufacturers[11].

Zeolite particles, as potential material for textile surfaces' modifications, are hydrated crystalline aluminosilicates consisting of three-dimensional frameworks of various numbers of SiO_4 and AlO_4 tetrahedra linked together in such a manner as to form channels and cavities within the size ranges 0.3 to 1.2 nm, filled by cations and water molecules[12]. Due to their particular characteristics, i.e. high thermal and hydrothermal stabilities, high surface area, exchangeable cations, and uniform internal channels and cavities, synthetic zeolites are exceedingly usable for numerous applications that are related to membrane separation, industrial ion exchange, adsorption, solid-acid catalysis processes, etc.[13,14] The field of surface modification of clothing textiles by zeolites has been poorly researched and have been focused mainly on cotton and polyester surfaces with enhanced thermal stabilities and flame-retardant properties[2,15,16]. The study realised by Carran et al. demonstrated an idea, where zeolite 5A was used on wool fabric for the creation of a superhydrophilic surface[17].

The aim of this work was to modify cotton fabric by zeolite particles using novel finishing-like procedures, and thus study the impacts of different operational parameters as well as auxiliaries on the degree of applied zeolites in order to determine better formulations along with the optimal technology. The emphasis was also given to the blocking abilities of modified textiles against harmful UV irradiation.

2.0. Materials and Methods

2.1. Materials

Trials were conducted using an industrially-bleached 100% cotton fabric (Co) in plain weave with a mass of 110 g/m^2, warp density of 41 threads/cm and weft density of 38 threads/cm, and warp/weft fineness of 20 tex. The zeolites used during this study were synthesised and supplied by SILKEM zeolites production, Inc., Kidričevo, Slovenia in the form of very fine powder; with average particle sizes of 3.45 μm, composed of 18.08 wt.% Na_2O, 29.24 wt.% Al_3O_3, 43.73 wt.% SiO_2, BET surface area of 204 m^2 g^{-1} and micro-pore volume of 0.095 cm^3 g^{-1}.

2.2. Zeolite Application Procedures

2.2.1. Exhaustion

Zeolites were applied on the cotton (Co) fabric according to the exhaustion procedure using a liquor-to-fabric weight ratio of 20:1 (160 mL of deionised water against 8 g of Co fabric) within a sealed stainless-steel pot of 200 cm^3 capacity, housed within the laboratory apparatus Labomat (W. Mathis) at a temperature of 60°C for 60 min. The six initial baths were differentially composed, namely: i) 3% owf (of weight of fabric) of zeolite (3% Z); ii) 3% owf of zeolite and 1 g L^{-1} of Meropan DPE - levelling agent (3% Z + lev.a); iii) 3% owf of zeolite, 1 g L^{-1} of Meropan DPE, 20 g L^{-1} of NaCl and 5 g L^{-1} of Na_2CO_3 (3% Z + aux.); iv) 6% owf of zeolite (6% Z); v) 6% owf of zeolite and 1 g L^{-1} of Meropan DPE (6% Z + lev.a); vi) 6% owf of zeolite, 1 g L^{-1} of Meropan DPE, 20 g L^{-1} of NaCl and 5 g L^{-1} of Na_2CO_3 (6% Z + aux.).

2.2.2. Pad-dry-thermofix

Zeolites (Z) in three concentrations (5, 10, or 30 g L^{-1}) with or without the additions of 40 g L^{-1} modified dimetylol dihydroxy ethylene urea (crosslinking agent - cr.a) were applied onto cotton fabric during the pad-dry-thermofix procedure. According to the previously optimised method, the samples were padded twice with a pad-liquor, by using a laboratory pad-mangle with 80% wet pick-up, dried for 3 min at a temperature of 100°C, and then treated at 150°C for 5 min.

All the zeolite-treated samples were washed at a temperature of 40°C for 30 min in a Labomat (W. Mathis) using a solution of 4 g L^{-1} of standard reference detergent without optical brighteners, and a liquor-to-fabric weight ratio of 50:1. Thereafter, the samples were rinsed in tap water, and finally dried at room temperature. After washing, the analytical measurements of the samples were carried-out.

2.3. Analytical Methods

X-ray powder diffraction (XRD) measurements of the zeolites were performed at room temperature in a XRD diffractometer PANalytical X'Pert PRO, using $CuK\alpha_1$ radiation (λ_1 = 1.5406 Å) over 2θ range of 5 to 50° over steps of 0.016° per 100 sec.

Scanning electron microscopy (SEM) was utilised for the morphological analysis of the zeolites, and for the surface morphologies of the zeolite-modified cotton fabrics. The samples were placed on an adhesive carbon band attached to the brass holder of a Zeiss Gemini Supra 35 VP Scanning Electron Microscope (Carl Zeiss NTS GmbH, Germany), and pictures taken.

Fourier transform infrared spectroscopy (FTIR) of the samples (zeolite powder and coated fabrics) was accomplished using a FTIR System Spectrum GX spectrophotometer (Perkin Elmer) with a Golden Gate ATR attachment and a diamond crystal. The absorbance spectra were obtained within the range of 4000-650 cm^{-1}, with 32 scans and a resolution of 4 cm^{-1}.

UV shielding capabilities of the zeolite-modified fabrics is usually expressed by the ultraviolet protection factor (UPF). UPF was calculated from the fabrics' UV-A and UV-B transmittance values, which were recorded according to the Australian/New Zealand Standard (AS/NZS 4399-1996) over an ultraviolet spectral region of 280-400 nm wavelengths using a solar-screen Cary 50 spectrophotometer (Varian), fitted with an integrated sphere accessory and a fabric holder accessory.

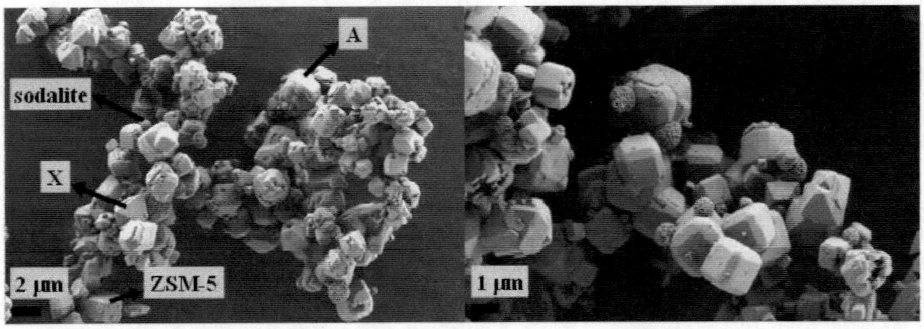

Fig. 1: SEM images of used zeolite at different magnifications

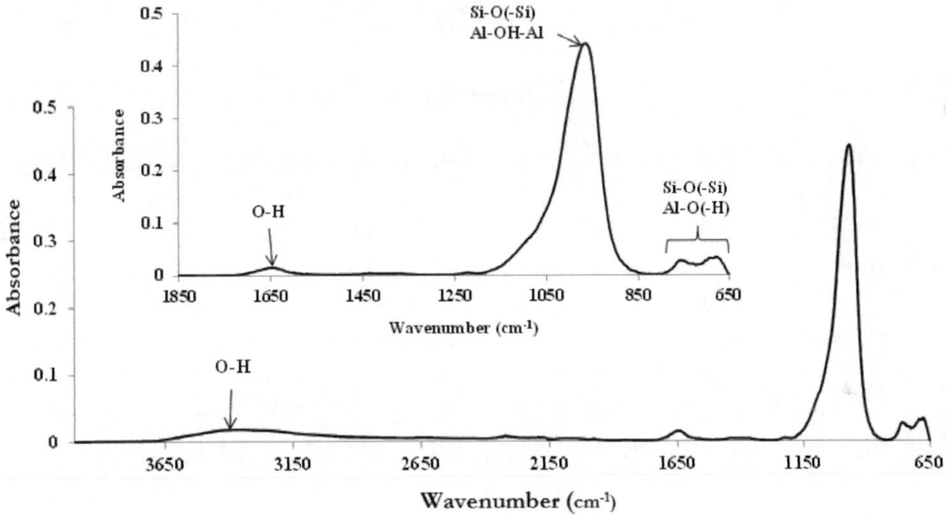

Fig. 2: FTIR spectrum of zeolite

3.0. Results and Discussions

3.1. Analysis of Zeolites

SEM images of the selected zeolite particles were taken with the intention of studying zeolite surface morphologies for the subsequent modifications of cotton fabric (Fig. 1). In order to further investigate the zeolite chemical characterisation on a molecular level, FTIR spectrum was scanned and presented in Fig. 2. Finally, the crystalline nature of zeolites was investigated by XRD analysis, as shown in Fig. 3.

Mixture of three different types of zeolites in sizes between 2 and 6 µm could be unequivocally perceived from the SEM micrographs in Fig. 1. Notably, the examined sample showed different ratio of cubic crystals of zeolite A in large amounts, followed by the octahedral crystals of zeolite X, and the pentasil building blocks of zeolite ZSM-5 in moderate amounts, which was also confirmed by X-ray powder diffraction measurement (Fig. 3). Furthermore, a high amount of sodalite was noticed as well as the presence of impurities, probably due to the industrial synthesis procedure.

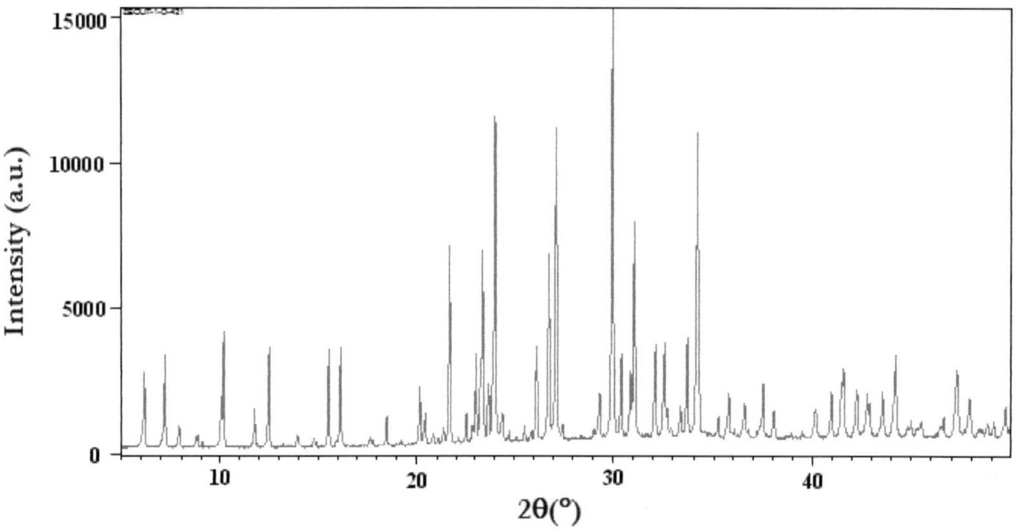

Fig. 3: X-ray diffractogram of zeolite

The FTIR spectrum of zeolite fine powder, as presented in Fig. 2, showed some characteristic bands, i.e. the most intensive band within the area of low frequency at a wavenumber of 970 cm^{-1} was due to Si–O(–Si) stretching vibrations and to bending modes of structural Al–OH–Al, the bands within the range of 680-800 cm^{-1} having been assigned to the 'ring' vibrations of internal oxygen bridges Si–O(–H) and Si–O–Si[18]. Finally, a moderate but wide peak could be observed at a wavenumber of ca. 3340 cm^{-1}, and also at 1650 cm^{-1} due to the O–H stretching and bending vibrations, respectively, associated with the presence of interlayer H_2O.

The XRD pattern in Figure 3 exhibited strong peaks characteristic for three different types of zeolites (A, X, and ZSM-5)[19,20] with high crystallinities as previously disclosed by SEM micrographs in Fig. 1.

3.2 Morphological and Chemical Characterisation of Zeolite-Modified Cotton

The surface morphologies of selected zeolite-modified/washed cotton fabrics were studied by SEM and the specific molecular vibrations on the surfaces by FTIR spectroscopy, with regard to different application techniques (exhaustion and pad-dry-thermofix) as well as the initial baths' compositions (concentrations of zeolites, additions of crosslinking agents and/or auxiliaries, temperatures, treatment times, etc.). The gained SEM micrographs are shown in Fig. 4 and infrared spectra in Fig. 5; the abbreviations for both figures are explained in Section 2.2.

The SEM micrograph in Figure 4(a) showed the clear and clean cellulose fibres' surface of untreated (reference) sample that was additionally coated with zeolites, thus changing the surfaces' morphologies - Figure 4(b-i). The most sufficient, but not ideal, coating was achieved by the pad-dry-thermofix procedure with or without the application of DMDHEU crosslinking agent. On the other hand, the exhaustion application technique - treatment at longer times in a bath at a temperature of 60°C resulted in lower particle distribution on the cotton fabric, depending upon the zeolite concentration (negligible effect at 3% owf of zeolites) as well as the chemical agents used.

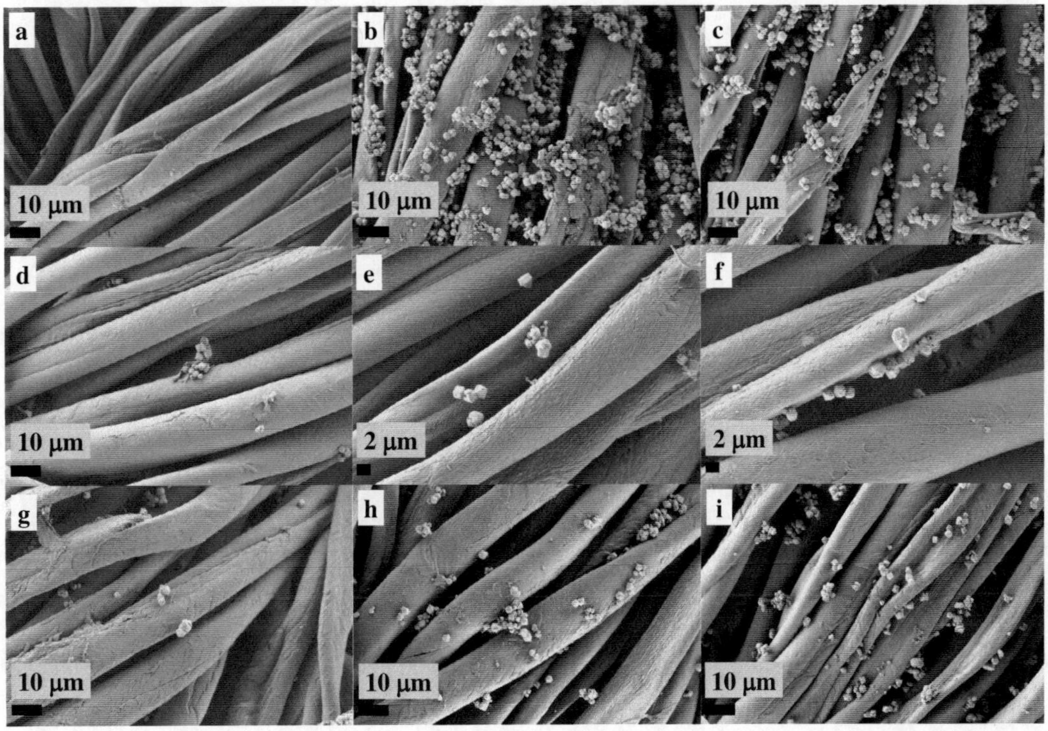

Fig. 4: Selected SEM images of: a) untreated cotton; and cotton modified by pad-dry-thermofix using: b) 30 g L^{-1} Z; c) 30 g L^{-1} Z + cr.a; and modified by exhaustion using: d) 3% Z; e) 3% Z + lev.a; f) 3% Z + aux.; g) 6% Z; h) 6% Z + lev.a; i) 6% Z + aux.;

Overall, it appears that both employed zeolite modification procedures had no deleterious effects on the morphology of the native cellulose fibre.

Fig. 5 depicts a FTIR pattern with typical peak positions for cotton including free hydroxyl groups at 3500-3200 cm^{-1}, water molecules at ~1645 cm^{-1}, O–H, C–O, C-H, and C–O–C vibrations within the glucose ring at 1500-800 cm^{-1}, as also fully interpreted.[1,21] Although, all the gained spectra were quite similar due to the overlapping of characteristic absorption peaks of cellulose (ref. in Fig. 5) and zeolites (Fig. 2), some absorption bands were evidently recognisable in the left diagrams for those zeolites modified Co fabrics (pad-dry-thermofix process), i.e. enlarged peak at 995 cm^{-1} corresponding to the stretching vibrations of the Si–O–Si bonds and around 893 cm^{-1} to the bending vibration modes of Al–OH–Al. Moreover, peaks at 1708 cm^{-1} and 1239 cm^{-1} were noted for samples treated with a combination of zeolite and crosslinking agents on account of the vibrations of –CO–NH–R bonds, which were formed between adjacent cellulose chains and DMDHEU during thermofixation. Meanwhile, minor changes between the FTIR spectra in right diagram could be observed, because of the lower amount of zeolite on the surface during the exhaustion procedure, as could also be expected from adequate SEM micrographs in Fig. 4.

3.3 UV-protective properties

The influences of the employed application techniques as well as the baths' compositions on the cotton fabrics' UV-blocking functionalities were followed by means of spectrophotometry.

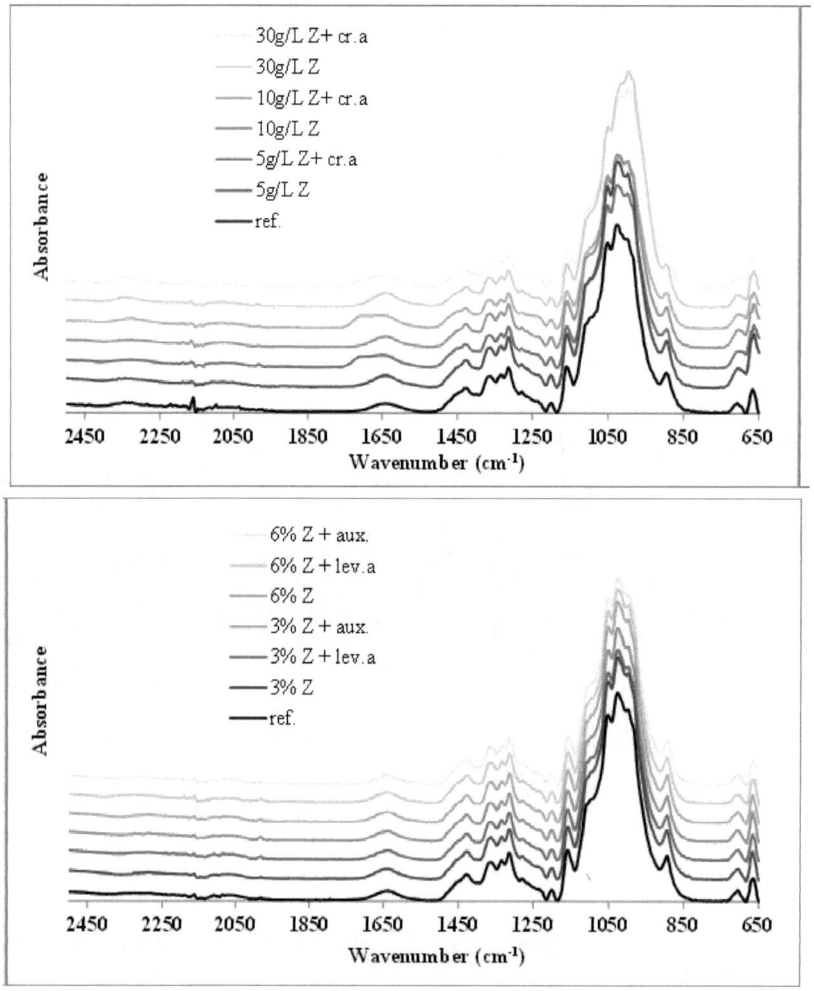

Fig. 5: Infrared spectra (4000-650 cm^{-1}) of cotton samples modified by zeolites according to the pad-dry-thermofix (above) and exhaustion (below) procedures.

The gained UPF results are presented in Fig. 6; the abbreviations are explained in Section 2.2.

It was evident from Figure 6 that the untreated cotton had UPF 3.4, indicating a non-ratable UV protection level. After the applications of 5, 10 and 30 g L^{-1} of zeolite particles during the pad-dry-thermofix procedure, the fabrics' UPFs exceedingly but non-linearly increased, as expected from literature, [15] i.e. up to values 22.4 (30 g L^{-1} of zeolite), and 26.3 (30 g L^{-1} of zeolite together with 40 g L^{-1} of DMDHEU), meaning good to very good UV protective properties with regard to ASTM guidelines for labelling the sun protective clothing (ASTM D 6603). The reduced UPF values were attained by the exhaustion procedure (up to 9.5) depending on the zeolite concentrations as well as the baths' compositions, which was in good analogy with the above-presented SEM results (Figure 4). Although, the reached UPFs of all the zeolite-modified/washed samples were lower than the value 55 (declared as UPF 50+), the gained results confirmed that zeolites impart a functionality to cellulose fabrics against the transmittance of harmful solar UV rays. Further employment of dyestuffs/printing pigments and finishing agents could also improve the UPFs, as is well-documented[1,15].

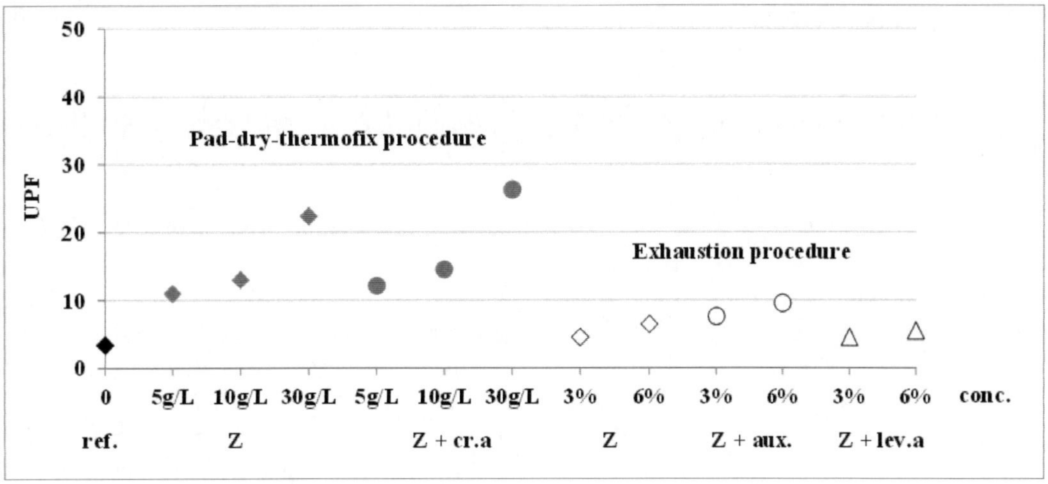

Fig. 6: Ultraviolet protective factor (UPF) of zeolite-modified cellulose samples.

4.0 Conclusions

- It could be concluded from the SEM, FTIR and XRD analyses of zeolite that the examined particles included mixture of the cubic crystals of zeolite A, the octahedral crystals of zeolite X, and the pentasil building blocks of zeolite ZSM-5 in different ratio.
- The selection of a proper application technique along with zeolites' concentrations and auxiliaries essentially influences the amount of deposited zeolites changing the fabrics' surface morfologies.
- Pad-dry-thermofix using 30 g L^{-1} of zeolites with or without crosslinking agent was the most appropriate procedure implying fabrics' good to very good UV-blocking properties.
- The selected exhaustion procedure, i.e. treatment of fabric 1 hour at 60°C with zeolites in different concentrations using different auxiliaries, was less suitable.

5.0 Acknowledgements

The research leading to these results has received funding from the Slovenian Research Agency (ARRS) under applied project no. L2-6776, and a programme group for Textile Chemistry P2-0118.

6.0 References

1. Fakin, D.; Veronovski, N; Ojstršek, A.; Božič, M.; Carbohydrate Polym., 2012, vol. 88(3), p. 992.
2. Monteiro, A.; Jarrais, B.; Rocha, I.M.; Pereira, C.; Pereira, M.F.R.; Freire, C.; Appl. Clay Sci., 2014, vol. 101, p. 304.
3. Ibrahim, N.A.; Eid, B.M.; Youssef, M.A.; Ibrahim, H.M.; Ameen, H.A.; Salah, A.M.; Carbohydrate Polym., 2013, vol. 97, p. 783.
4. Montazer, M.; Alimohammadi, F.; Shamei, A.; Rahimi, M.K.; Carbohydrate Polym., 2012, vol. 87(2), p. 1706.
5. Sivakumar, P.M.; Balaji, S.; Prabhawathi, V.; Neelakandan, R.; Manoharan, P.T.; Doble, M.; Carbohydrate Polym., 2010, vol. 79(3), p. 717.
6. Veronovski, N.; Hribernik, S.; Andreozzi, P.; Sfiligoj Smole M.; Fibers and Polym., 2009, vol. 10(5), p. 716.
7. Tsafack, M.J.; Levalois-Grützmacher, J.; Surf. Coat. Technol., 2007, vol. 201(12), p. 5789.

8. Textor, T.; Mahltig, B.; Appl. Surf. Sci., 2010, vol. 256(6), p. 1668.
9. Goncalves, G; Marques, P.A.A.P.; Pinto, R.J.B.; Trindade, T.; Neto, C.P.; Composites Sci. Technol., 2009, vol. 69(7-8), p. 1051.
10. Liu, J.; Wang, Q.; Fan, X.R.; J. of Sol-Gel Sci. Technol., 2012, vol. 62(3), p. 338.
11. Ojstršek, A.; Stana Kleinschek, K.; Fakin, D.; Surf. Coat. Technol., 2013, vol. 226, p. 68.
12. Liu, S.; Cao, X.; Li, L.; Ji, Y.; Xiao, F.S.; Colloids and Surfaces A: Physicochemical and Engineering Aspects, 2008, vol. 318(1-3), p. 269.
13. Meng, X.; Nawaz, F.; Xiao, F.S.; Nano. Today, 2009, vol. 4(4), p. 292.
14. Bastani, D.; Esmaeili, N.; Asadollahi, M.; A review, J. Ind. Eng. Chem., 2013, vol. 19(2), p. 375.
15. Grancarić, A.M.; Tarbuk, A.; Kovaček, I.; Chem. Ind. Chem. Eng. Quarterly, 2009, vol. 15(4), p. 203.
16. Grancarić, A.M.; Tarbuk, A.; McCall, D.; Polimeri, 2007, vol. 28(4), p. 219.
17. Carran, R.S.; Ghosh, A.; Dyer, J.M.; Appl. Surf. Sci., 2013, vol. 287, p. 467.
18. Mozgava, W.; J. Mol. Structure, 2001, vol. 596(1-3), p. 129.
19. Liu, S.; Cao, X.; Li, L.; Li, C.; Ji, Y.; Xiao, F.S.; Colloids and Surfaces A: Physicochemical and Engineering Aspects, 2008, vol. 318(1-3), p. 269.
20. Cheng, Y.; Li, J.-S.; Wang, L.-J. Sun, X.-Y.; Liu, X.-D.; Separation and Purification Technology, 2006, vol. 51, p. 210.
21. Galkina, O.L.; Sycheva, A.; Blagodatskiy, A.; Kaptay, G.; Katanaev, V.L.; Seisenbaeva, G.A.; Kessler, V.G.; Agafonov, A.V.; Surf. Coat. Technol., 2014, vol. 253, p. 171.

Residual Stress Investigations on DLC and TiN Films Deposited onto A390 Aluminum Alloy for Industrial Applications

I. Peter*, M. Rosso, C. Castella,
Politecnico di Torino, Department of Applied Science and Technology,
Corso Duca degli Abruzzi 24, Torino, Italy.

A. Bendavid
CSIRO Manufacturing Flagship, Sydney, NSW, Australia.
*Email: ildiko.peter@polito.it

Abstract

Hypereutectic Al-Si alloys have attractive properties making them appropriate materials for automotive and aerospace components production. For engine components, fatigue and wear resistance are fundamental properties because they are directly associated to their failure. The tribological properties of A390 Aluminum alloys have been enhanced through the deposition of diamond-like carbon (DLC) coating by plasma-enhanced chemical vapor deposition (PECVD) technique and by titanium nitride (TiN) coating deposited by filtered cathodic vacuum arc deposition (FCVAD) technique. In many cases unexpected failure has occurred which can be attributed to the presence of residual stresses, generated during manufacturing or successive machining procedures, resulting in considerably reduction in the life of the component. In this paper X-ray techniques and d $\sin^2 \psi$ method is applied to determine the residual stress of the Al substrate and of the coated surfaces. Comparison between the levels of the residual stress is performed and correlated the findings to the materials microstructural and tribological behavior.

1.0. Introduction

Selection of materials for an application is dependent on their properties and constitutes a very important issue in any industrial application. The driving force for the introduction of light metals and their alloys in automotive and aeronautical industries is due to their high strength-to-weight ratio, good formability, excellent combination of castability, mechanical properties, and excellent corrosion resistance, which make them very suitable for a number of applications. The use of Al–Si alloys as an alternative solution for Fe-based products has been promoted, generally, with a view to decrease the total weight of the components and taking into account environment issues. However, despite of their attractive properties and due to a lack of competitiveness concerning their tribological properties compared to Fe-based metals, industrial application are not utilized as well as expected. Generally, lightweight metals and their alloys are not strong enough to be used in extremely high loading conditions or in wear resistant applications: as a rule, the tribological properties of Al alloys are poor, apart from the high Si content alloys[1–7].

Al has higher corrosion resistance than Fe and steel, but it's resistance can be further enhanced in case of decorative materials by surface modification through the deposition of corrosion resistant film[8]. Over the years, surface modification of Al–Si alloys aiming to improve their mechanical performance, has received more attention from the research community and industry. Plasma spray coatings, wear resistant plating, friction stir processing, chemical conversion, solid lubricant coatings, anodizing, chemical, physical vapor depositions, alloying by electron beam, laser and arc or gas plasma heat sources can effectively improve the surface properties of lightweight alloys and increase their durability[9–12].

Generally, hypereutectic Al-Si alloys are less accessible to grain refinement through conventionally grain refiners and other methods have to be considered to improve their tribological properties. Chemical eutectic modification involves the development of porosity and hot tearing with poor surface finishing properties. The use of mechanical or electromagnetic stirring and ultrasonic irradiation have been extensively explored[1,2] but some effects on the eutectic phase remains unclear and the origin of primary-phase modification has been discussed.[12] From industrial point of view, the deposition of some protective coatings are considered very useful to improve the tribological properties of aluminum alloys. Indeed in recent years, diamond-like carbon (DLC) and titanium nitride (TiN) coatings have been widely studied and used as wear resistance coatings, for their low coefficient of friction and high wear resistance. DLC is a metastable structure of amorphous carbon, consisting of a network of sp^3 and sp^2 bonded carbon atoms[13–14].

In the automobile sector this coating is extensively used as a protective coating for components such as piston, valves and camshafts[15]. The adhesion of DLC on AA5052 aluminum alloy has been reported to be good when deposited by plasma immersion ion implantation, and suitable method to increase the alloy's corrosion resistance of 100 times.[16] It has been also demonstrated that the aluminum alloys wear resistance can be increased by the deposition of DLC-based coatings, with multi-layered structures, deposited by a combined arc ion plating and magnetron sputtering process.[17] The wear resistance and the surface hardness of 2024 aluminum alloy can be increased by the deposition of DLC coating[15]. DLC coating and doped DLC with low concentration of Cr are widely investigated in the aim of reducing the friction behavior and improving the wear properties of aluminum alloys[18]. Another interesting wear resistance coating commonly used is TiN, which is predominately deposited by physical vapour deposition (PVD) or chemical vapour deposition (CVD), but since these two deposition techniques present a low deposition rate[19–23], plasma spraying and high-velocity oxy fuel (HVOF) technologies have been studied[21–26]. Recently, it has been demonstrated that the tensile properties of a 7075-T6 aluminum alloy, coated with TiN deposited by PVD are drastically reduced due to the high temperature deposition process resulted in an enlargement of the strengthening precipitates. With the use of T6 heat treatment most of the properties can be restored[27]. In many cases, where unexpected failure has occurred due to the presence of residual stresses which resulted in the reduction of the life of the component. Both material and film deposition processes can inducee residual stresses, which can be a critical cause for failure in some applications. Some of the causes of stress formation are: thermal mismatch, structural and mechanical properties between the substrate and the coating layer. Heat treatable Al alloys are subject to severe thermal gradients when quenched from the solution heat treatment temperature; these gradients originates a non-homogeneous plastic flow and can cause deformation and residual stresses in the final components. The nature of the total residual stress in the coated material can be determined by the criteria identified by Stokes and Looney[28]. The major effect is originating from the cooling stresses associated to the relative values of their coefficient of thermal expansion (α) and based on its value the stress can be classified[29].

In addition, residual stresses can be generated in the machined surface layer and after any deposition procedure, because of the expansion of the lattice and precipitation of nitrides. With modern analytical and computational techniques it is possible to evaluate the stresses to which a component is subjected during usage. A large number of residual stress measurement techniques are in use, some of them are destructive, while others non-destructive. The most common methods to investigate residual stress are curvature measurements, hole drilling, compliance methods and X-ray/Neutron/Electron diffraction methods. Diffraction methods of residual stress determination measure the angles at which the maximum diffracted intensity occurs when a crystalline sample is subjected to X-rays. The angles allows the determination of the interplanar spacing "d" of the diffraction planes using Bragg's law: if residual stresses are present, then the spacing "d" will be different with respect to the unstressed state and the variation is proportional to the degree of the residual stress[30].

In this paper, investigations un-coated Al alloy substrate, DLC, TiN coated Al alloy substrates, and Al alloy substrate after the removal of the DLC and TiN coatings have been investigated. Residual stress has been measured experimentally and correlated to the morphological and tribological behavior of the samples.

2.0. Materials and Methods

Some samples made of A390 aluminum alloy (Si 18.4%, Cu 4%, Fe 0.23%, Mg 0.57%, Mn 0.07%, Ni 0.02%, Al balance) have been prepared by a standard metallographic polishing technique, ultrasonically cleaned in acetone and finally dried in air prior to deposition. TiN protective coating has been deposited by filtered cathodic vacuum arc deposition (FCVAD) technique. DLC coatings were deposited using plasma-enhanced chemical vapor deposition method (PECVD). The residual stress has been determined by the d $\sin^2 \psi$ method.

DLC coating was deposited using a radio frequency (RF) PECVD technique. The chamber was evacuated to a pressure of approximately 2×10^{-5} mbar. Prior to deposition, the samples were *in situ* sputter cleaned for 10 min in argon plasma operated at 25 mTorr with the argon flow rate set at 10 sccm at 200 W power, and the self-bias was measured to be 630 V. To improve the adhesion, a thin layer of hydrogenated amorphous silicon carbide (a: SiCH) was first deposited onto the substrate using tetramethylsilane (TMS) as precursor. The interlayer deposition was performed at 25mTorr with the TMS flow rate was set at 85sccm at 200 W power for 5 min. For the deposition of the DLC coatings, acetylene (C_2H_2) gas was introduced into the deposition chamber at a flow rate of 100 sccm, and for 10min the pressure and power were set to 50 mT and 200 W respectively[31].

The vacuum chamber used for the TiN coating was equipped with a diffusion pump reaching a base pressure of 10^{-3} Pa and during the deposition a variable throttle valve was used to control the pressure. The arc current used was 120 A and in these conditions the produced Ti^{+} current measured at the substrate was 0.8 A. The arc source uses a 55mm diameter Ti (99.9% purity) cathode. The substrate was heated to 300°C by radiative heating and the temperature was monitored by a k-type thermocouple. The substrate holder was also connected to a d.c. power supply in order to bias the sample during film growth. In this investigation, the substrate bias was set at -100 V. Research grade nitrogen gas was introduced into the system through mass flow controllers. The deposition pressure was adjusted to 2.5 mTorr. Prior to deposition, the samples were in situ sputter cleaned for 10min using anode layer ion source at 0.8 mTorr argon pressure. To improve the adhesion an interlayer of Ti was deposited prior to the TiN deposition[32].

Ball-on-disk test was used to evaluate the tribological properties of the two coatings. A WC ball, with a diameter of 6 mm, has been used as friction counterpart, while the wear track was set to be 12 mm from the center of the sample. An applied load of 5N, a speed of 11 m/s and a sliding distance of 500 m have been set for the tests.

X-ray techniques (X-ray, PANanalytical tool) and d $\sin^2 \psi$ method (ψ being the tilt angle) has been implemented on the (3 1 1) diffraction plane of Al alloy substrate prior, after coating and following the removal of both coating layers for ψ ranging from 0 to 33.21°. For the TiN coating the (1 1 2) diffraction plane and ψ ranging from -14.18° to +14.18° was considered. Comparison between the different conditions have been performed by evaluating the stress measured in the un-stressed Al alloy, before coating and after coating and following the removal of the DLC and TiN coating layer.

3.0. Results and Discussions

3.1. Wear test

The coefficient of friction vs sliding distance curves of the DLC and TiN coatings are shown in Fig. 1. The DLC coating has presented a steady coefficient of friction (about 0.17) for the entire test while the TiN coating has showed a completely different behaviour. Indeed TiN tests show after a sliding distance of about 20m, a sharp increase of the coefficient of friction, due to the failure of the protective coating.

The failure can be attributed to the presence of some defects on the substrate just before the deposition process, as shown in Fig.2(a), which have gave rise to defects on the coating surface, as can be observed in Fig. 2(b). These defects have resulted an increase in the wear rate, due to increase of the occurrence of stress concentrations around them and so consequently the failure of the protective coating at the earlier stage of the wear test.

Finally, the coefficient of friction of TiN has reached a value of about 0.6-0.7, which is similar to that of the A390 aluminum alloy used as a substrate[7]. DLC coating resulted an enhancement of the tribological behaviour of the A390 aluminum alloy, while the presence of defects in the TiN coating has had a deleterious effect on its wear resistance.

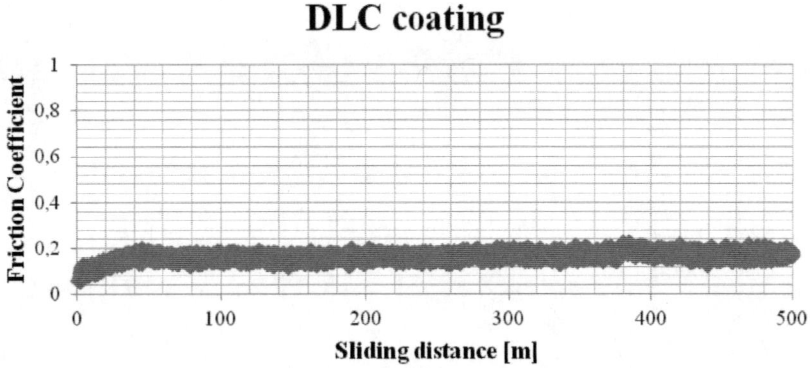

a

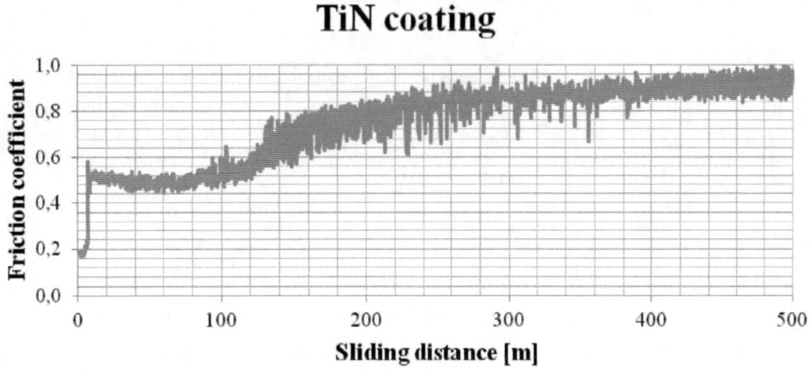

b

Fig. 1: Friction coefficient versus sliding distance for: **(a)** DLC coating and **(b)** TiN coating

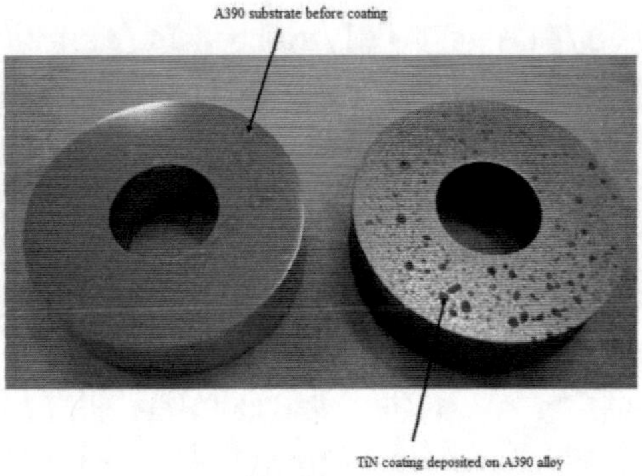

Fig. 2: Defects observed in the substrate and in the TiN coating

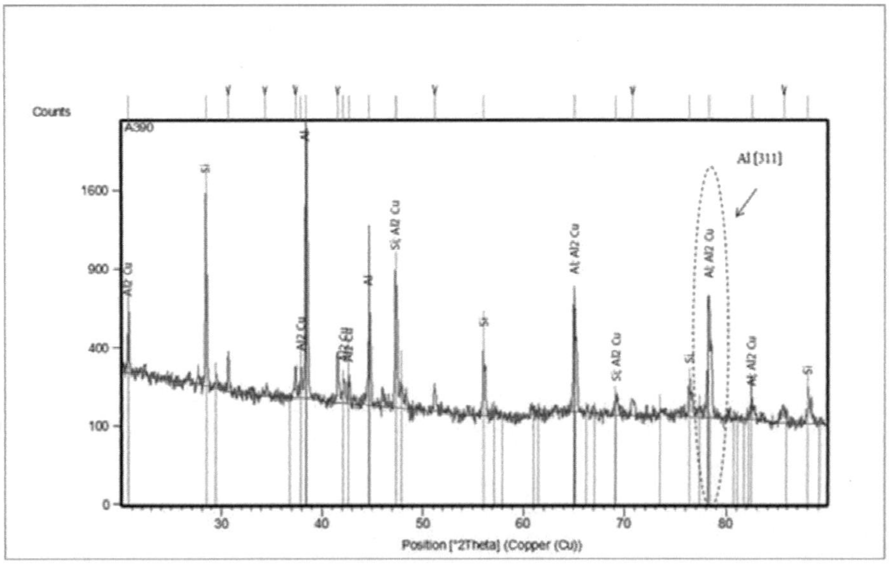

Fig. 3: X-ray diffraction pattern for the A390 alloy sample prior to coating procedure

3.2. X-ray diffraction and sin2 ψ measurements

The diffraction angle, 2Θ, is measured experimentally and then the lattice spacing (d) is calculated from the diffraction angle and the known X-ray wavelength (λ) using Bragg's Law (n λ = 2 d sinΘ, where n=1, 2, 3…). When residual stresses are present in the material, the d-spacing will be different compared to the un-stressed state. The difference is proportional to the level of the residual stress. Once the d-spacing values are identified, they can be plotted versus sin2 ψ. The plot of d vs. sin2 ψ is a straight line, for these cases, which slope is proportional to the stress present in the Al alloy analysed, using the equation [32]:

$$\sigma = (E/(1+\nu)) \bullet (1/do) \bullet (\delta d/\delta sin2\psi)$$

where: do = un-stressed interplanar lattice spacing;
d = stressed interplanar lattice spacing;
E = elastic modulus for a certain plane with Miller index;
ν = the Poisson modulus associated to the same crystallographic plane;
ψ = tilt angle

Fig. 3 shows a typical X-ray pattern of the un-coated A390 alloy sample. The arrow indicates the (3 1 1) diffraction peak positioned at 2Θ angle of 78.229° that was selected for the residual stress measurement for all the three samples in order to compare the stress level before and after coating application.

When the sample is tilted with respect to the in-coming X-ray new grains will diffract and the orientation of the diffraction planes is practically perpendicular to the stress direction. As a result of the tilt, the interplanar spacing decreases and 2Θ increases in the case of tension, while the interplanar spacing increases and 2Θ decreases in case of compression (as a result of Poisson's effect).

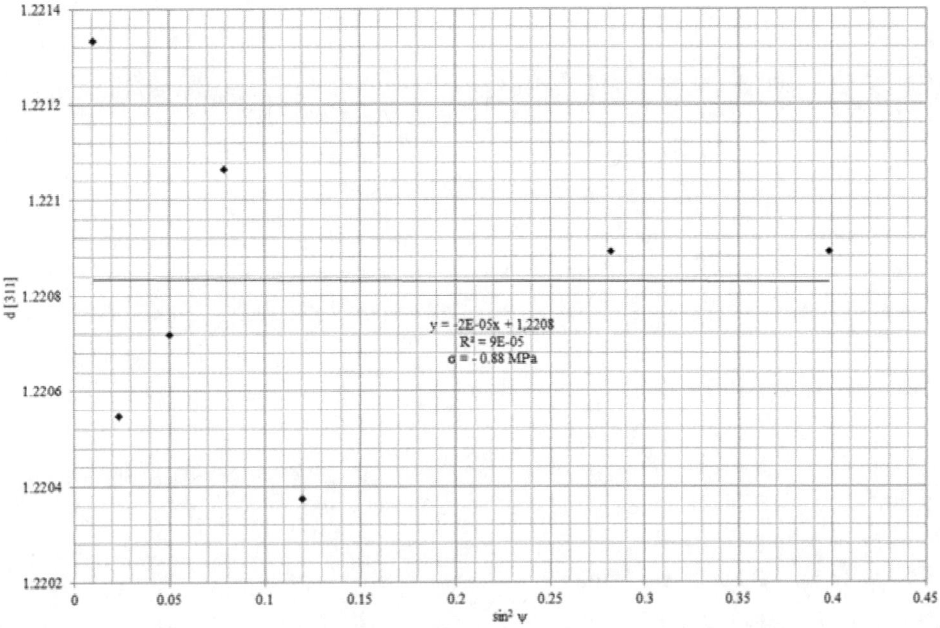

Fig. 4: A linear dependence of d (3 1 1) upon $\sin^2 \psi$ for A390 alloy samples before coating

The interplanar spacing acts as a strain gauge. Any type of stress has an effect on the variation of the interplanar spacing, which has a very low level of change. For the TiN layer the (1 1 2) diffraction peak positioned at 2Θ angle of 36.5° was selected for the residual stress measurement which is free from any peak interference.

The size of the substrate samples used for stress measurements were 10 x 10 x 10 mm. Layers having a target thickness of about 5 μm of DLC coating and of about 6 μm of TiN coating were progressively removed in order to access in-depth residual stress measurements. Removal of DLC and TiN coating layers were carried out by polishing procedures carried out carefully to avoid additional mechanical solicitations in the samples.

As reported in Fig. 4 the A390 aluminum alloy substrate presents a very low compressive residual stress (-0.88 MPa), meaning that, before deposition, the substrate was not under tensile stress, the fabrication and the subsequent finishing and preparation procedures did not alter significantly the tensile behaviour of the alloy.

The PECVD deposition process takes place with a high deposition rate, but at low temperatures. In this way, the temperature difference between the substrate and the coating layer is significantly low. Additionally, the relaxation of carbon atoms and ions adsorbed onto the surface favors a stress release[34].

Following elimination of the coating layer, as illustrated in Fig. 5, a low level of residual stress is present (+ 17.68 MPa). Generally, dense TiN films have compressive stress in the diffusion region, due to the expansion of the lattice and precipitation of nitrides.

The magnitude of residual stresses generated in the TiN layer is high compressive stresses (-1.38 GPa) as reported in Fig. 6 and also after the removal of the film the level of the tensional state of the substrate (Fig. 7) was high.

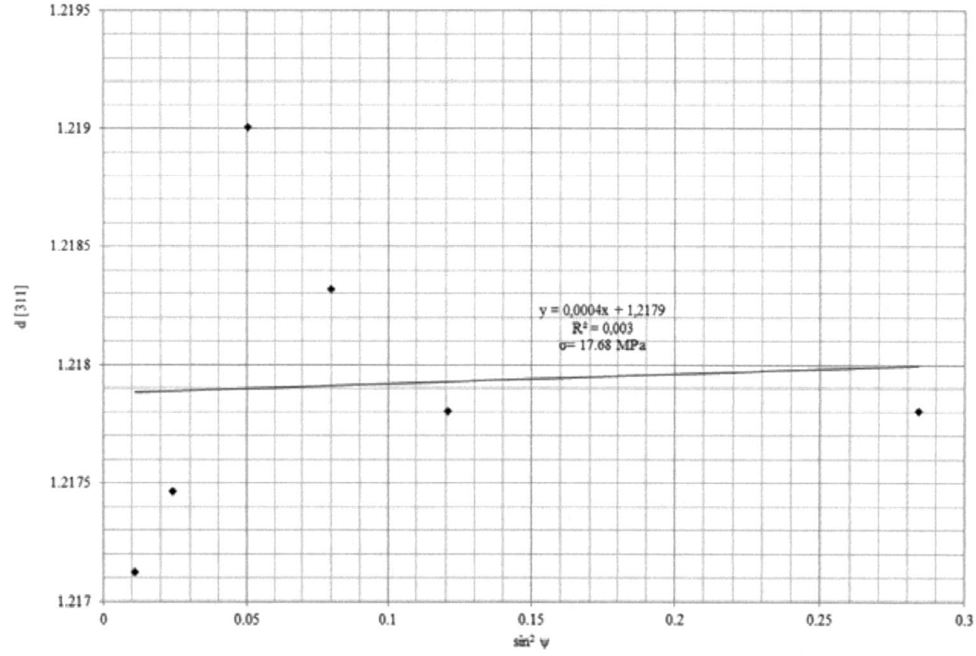

Fig. 5: A linear dependence of d (3 1 1) upon $\sin^2 \psi$ for A390 alloy samples after removal of DLC coating

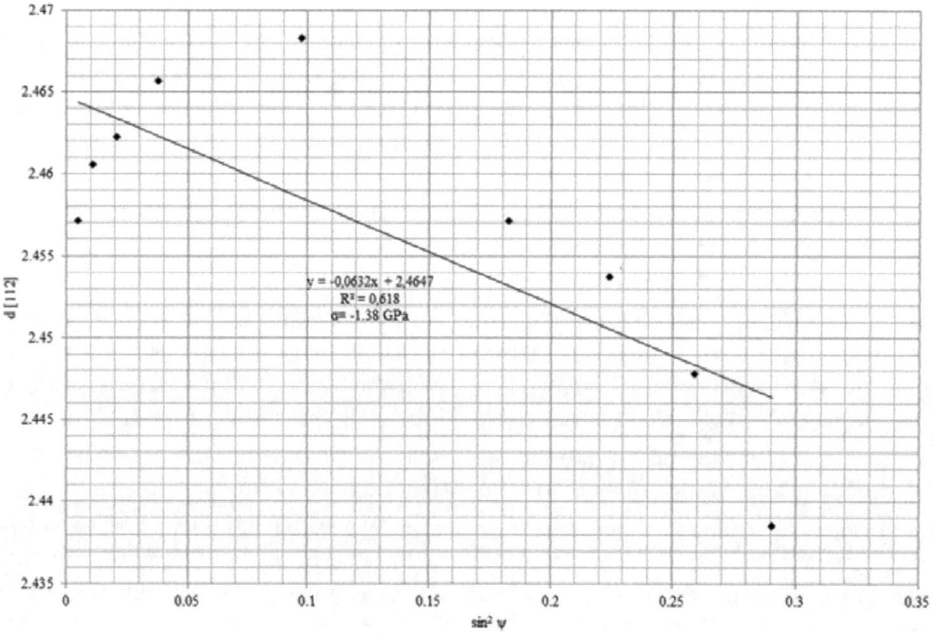

Fig. 6: A linear dependence of d (1 1 2) upon $\sin^2 \psi$ for A390 alloy TiN coated samples

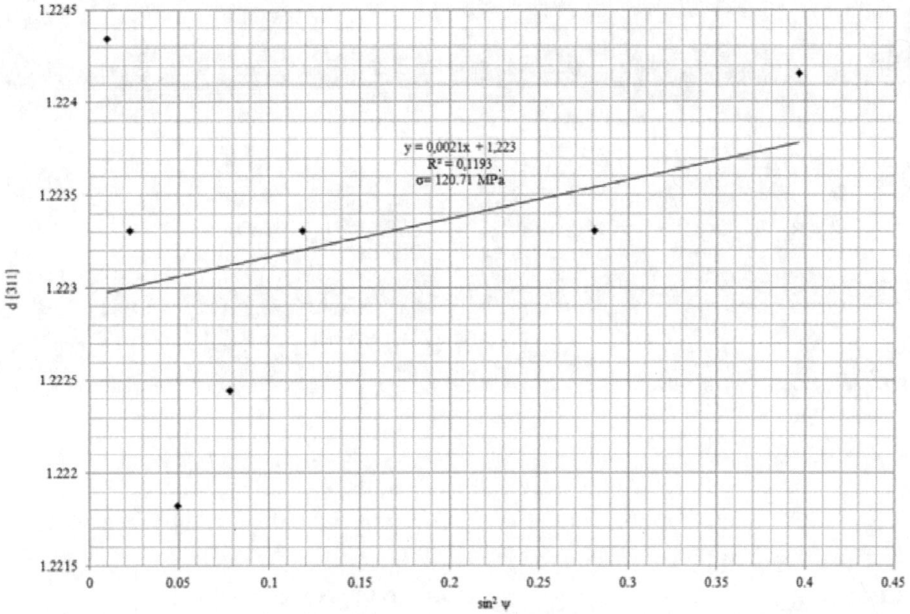

Fig. 7: A linear dependence of d (3 1 1) upon $\sin^2 \psi$ for A390 alloy samples after removal of TiN coating

DLC is amorphous and, measurement of residual stress cannot be carried out from diffraction pattern of the DLC film. In this investigation it has been proposed to measure the residual stress of DLC coated sample using the previously selected diffraction peak of Al (311) positioned at 2Θ angle of 78.229° of the substrate, in order to try to compare the results obtained and the level of the stress previously achieved. In this way, an approximation of the magnitude of the stress induced by the coating procedure and by the mechanical polishing has been realized: DLC coating layer show a lower level of residual stress (13.26 MPa) considering the deviation of the interplanar spacing relative to Al compared to the Al alloy substrate polished and following the removal of the coating layer. According to these results, the difference corresponds to 4.42 MPa and can be attributed to the preparing procedure for the removal of the coating.

The results obtained in this investigation are in good agreement with the results obtained during microstructural analysis and the tribological properties evaluation of the two protective coatings carried out previously[21].

The DLC film exhibit a good adhesion to the A390 aluminum alloy substrate and no obvious cracks or delaminations were observed as reported in Fig. 8. In the case of TiN coating layer there are some imperfections due to the presence of some defects, as shown in Fig. 2, on the substrate prior to the deposition.

As results of the defects on the substrate, development of some cracks are visible on the micrographs in Fig. 8. Without these defects, and the material is in compression, there are no other evident cause to the material's failure. The results obtained in this investigation are in agreement to the fact that TiN coating has been removed during the ball-on-disk test, just after about 20 m, while DLC coating does not present any wear tracks after about 500 m[35].

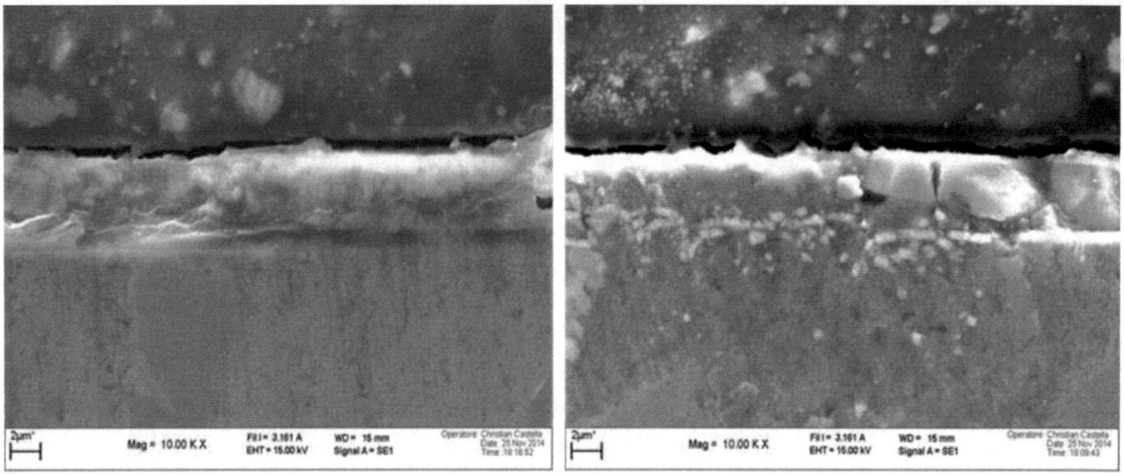

Fig. 8: SEM micrographs of **(a)** PECVD DLC coating **(b)** FCVAD TiN coating

4.0. Conclusion

In this paper, the residual stress of an un-coated Al alloy substrate, DLC and TiN coated Al alloy substrate and on the Al alloys substrate after the removal of the coating layer were measured.

- The analysis demonstrate the effect of manufacturing/deposition processes on the residual stress and highlighting is the importance of controlling the design and the production route of the metallic materials.
- The residual stress estimated by diffraction method for the samples are in line with the morphological and tribological performances.
- The TiN layer is in high compression, but failure occurs during tribological test: the layer was removed just after about 20 m due to presence of some defects on the coating layer originated from the initial state of the Al alloys substrate. On the other hand, DLC coating revealed high level of adhesion and does not present any wear tracks after about 500 m and no significant stress was measured.

5.0. References

1. ASM Handbook ASM International ed, Materials Park, OH, 1990, vol. 2.
2. ASM Specialty Handbook, J. E. DAVIS ed, Materials Park, OH, 1993.
3. Polmear, I.J.; Metallurgy of Light Alloys, John Wiley & Sons ed, Toronto, 1996.
4. Aluminum: properties and physical metallurgy,' J. E. Hatch ed, Materials Park, OH, 1984.
5. I. Peter, M. Rosso, C. Bivol, ' Microstructure and mechanical behaviour of Al-based alloy obtained by liquid forging technique,' Metallurgia International (2009), 14 15-19.
6. Rosso, M.; Peter, I.; Int. J. Microstructure and Mater. Properties, 2013, vol. 8(1-2), p. 113.
7. Mahmoud, T.S.; Surf. Coat. Technol., 2013, vol. 228, p. 209.
8. Ye, H.; J. Mater. Eng. Perf., 2003, vol. 12, p. 288.
9. Elmadagli, M.; Perry, T.; Alpas, A.T.; Wear, 2007, vol. 262, p. 79.
10. Ishii, H.; Furuyama, O.; Tanaka, S.; Metal Finishing, 1993, vol. 91, p. 7.

11. Zhang, F.; Yan, M.; J. Mater. Sci. Technol., 2014, p. 1.
12. Uenishi, K.; Kobayashi, K.F.; Intermet., 1999, vol. 7, p. 553.
13. Ni, W.; Cheng, Y.T. ; Weiner, A.M.; Perry, T.A.; Surf. Coat. Technol., *2006,* vol. 201, p. 3229.
14. Zou, C.W.; Wang, H.J.; Feng, L.; Xue, S.W.; Appl. Surf. Sci., 2013, vol. 286, p. 137.
15. Zou, Y.S.; Zhou, K.; Wu, Y.F.; Yang, H.; Cang, K.; Song, G.H.; Vacc., 2012, vol. 86, p. 1141.
16. Rangel R.C.C.; Souza M.E.P.; Schreiner W.H.; Freire C.M.A.; Rangel E.C.; Cruz N.C.; Surf. Coat. Technol., 2010, p. 3022.
17. Wang, L.; Wan, S.; Wang, S.C.; Wood, R.J.K.; Xue, Q.J.; Tribol. Lett., 2010, vol. 38, p. 155.
18. Zou, C.W.; Wang, H.J.; Feng, L.; Xue, S.W.; Appl. Surf. Sci., 2013, vol. 286, p. 137.
19. Stoiber, M.; Panzenb, C.K.M.; Mitterer, C.; Lugmair, C.; Surf. Coat. Technol., 2001, vol. 142, p. 117.
20. Motte, P.; Proust, M.; Torres, J.; Gobil, Y.; Morand, Y.; Palleau, J.; Pantel, R.; Microelectroni Eng., 2000, vol. 50, p. 369.
21. Jing, M.A.; Yan, D.Q.; Hu, J.W.; Zhang, X.; Li, Y.; Trans. Nonferrous Met. Soc. China, 2013, vol. 23, p. 1011.
22. Chatterjee, S.; Chandrashekhar, S.; Sudarshan, T.S.; J. Mater. Sci., 1992, vol. 27, p. 3409.
23. Braza, J.F.; Sudarshan, T.S.; Mater. Sci. Techonol., 1992, vol. 8, p. 574.
24. Feng, W.R.; Yan, D.R.; He, J.N.; Zhang, G.L.; Chen, G.L.; Gu, W.C.; Yang, S.Z.; Appl. Surf. Sci., 2005, vol. 243, p. 204.
25. Xiao, L.S.; Yan, D.R.; He, J.N.; Zhu, L.; Dong, Y.C.; Zhang, J.X.; Li, X.Z.; Appl. Surf. Sci., 2007, vol. 253, p. 7535.
26. Mao, Z.P.; Ma, J.; Sun, B.D.; Wang, J.; J. Coat. Technol. Res., 2009, vol. 6, p. 243.
27. Oskouei, R.H.; Ibrahim, R.N.; Surf. Coat. Technol., 2011, vol. 205, p. 3967.
28. Stokes, J.; Looney, L.; Proc. 1st Int. Therm. Spray Conf., 2000.
29. Oladijo, O.P.; Venter, A.M.; Cornish, L.A.; Sacks, N.; Surf. Coat. Technol., 2012, vol. 206, p. 4725.
30. Withers, P.J.; Bhadeshia, H.K.D.H.; Mater. Sci. Technol., 2000, vol. 17, p. 355.
31. Bendavid, A.; Martin, P.J.; J. Australian Cer. Soc., 2014, vol. 50(1), p. 86.
32. Martin, P.J.; Bendavid, A.; Thin Solid Films, 2001, vol. 394(1-2), p. 1.
33. Noyan, I.C.; Cohen, J.B.; Residual Stress Measurement by Diffraction and Interpretation, B. Ilschner and N. J. Grant ed, 1987.
34. Corbella, C.; Rubio-Roy, M.; Bertran, E.; Polo, M.C.; Pascual, E.; Andújar, J.L.; Diamond & Related Materials (2009) 18 1035–1038.
35. Peter, I.; Castella, C.; Rosso, M.; Bendavid, A.; Proc. Int. Therm. Spray Conf. Exposition, Long Beach California.

Fabrication of Zeolite-Coated Fibre-Forming Polymers

A. Ojstršek*,
University of Maribor, Faculty of Mechanical Engineering,
Institute for Engineering Materials and Design,
Smetanova 17, 2000 Maribor, Slovenia.

T. Fakin,
National Institute of Chemistry, Hajdrihova 19,
1000 Ljubljana, Slovenia

K. Stana Kleinschek, D. Fakin*
University of Maribor, Faculty of Mechanical Engineering,
Institute for Engineering Materials and Design,
Smetanova 17, 2000 Maribor, Slovenia.
*Email: alenka.ojstrsek@um.si, darinka.fakin@um.si;

Abstract

The aim of this research was to characterise four different types of zeolites in the forms of very fine powders for further application on the surface of polyamide knitted fabric using a novel industrially-acceptable exhaustion procedure with the intention of changing the fabrics' water absorption abilities. The obtained surface morphological observations and identifications of specific molecular vibrations' results of zeolite-coated polyamide knitted fabrics proven the suitability of employed exhaustion procedure regarding the types of zeolite used as well as the compositions of the treatment baths. In addition, the biggest changes in the water contact angle measurement results 8.6-20.3° - imposingly hydrophilic surfaces were obtained when using slightly acidic (pH 5.5 - 6.5) treatment baths composed of zeolites (irrespective of which type) in combination with pre-optimised auxiliaries. On the other hand, the applications of alkaline aqueous dispersions of ZSM-5 and 13X zeolites enlarged the contact angles up to 132.8° and 127.1°, respectively. In additionally, the optical properties were changed in both, UV and visible regions as well as visually perceivable changes in lightness (and somewhat less in chroma) occurring in regard to the types of zeolites and compositions of the treatment baths.

1.0. Introduction

Even from ancient times, humans have used various physical and chemical methods to establish aesthetic and functional properties to fibres[1]. Two broad categories of function have become critical in the textile production of synthetic textiles, namely the comfort and the protective functions; regardless of environment and activities. Polyamide fibres (PA) are linear polymers, the monomeric units of which are joined through secondary amide (-NHCO-) linkage[2]. This group of synthetic polycondensate fibres is important from a textile-technical and economic point of view, due to its favourable properties such as superior strength characteristics (abrasion resistant), good bacterial resistance, low swelling in aqueous media, excellent dimensional stability when washed, etc[3]. However, polyamide is relatively hydrophobic in character with the moisture regain being only about four percent[4,5]. Because of this feature, the capability of polyamide textiles to effectively transport moisture and heat away from the body are limited, and also a static charge on clothing and outfits can be created. Functional surface coatings that address thermal and/or sensorial comfort, which relates to skin contact sensation, have been increasingly important over recent years[1].

In addition to various nanoparticles that are used for modifying fibre-forming polymers, crystalline aluminosilicate three-dimensional microporous particles of numerous types have attracted augmented interest for such kinds of application on account of their superior heat resistance, high surface area, exchangeable cations, and high chemical stability[6–8]. Because they are non-toxic and non-absorbable by humans, zeolite particles are highly promising for the fabrication of multifunctional textiles that are constantly in direct contact with the skin. Grancarić et al., through several studies, demonstrated an idea as to how the application of tribomechanically activated natural zeolites could lead to the enhanced antimicrobial and UV protection capabilities of cotton and polyester fabrics[9,10]. Carran et al. concluded that surface treatment of wool fabrics with zeolite molecular sieve 5A leads to varied changes in the physicochemical properties of wool[11]. Monteiro and co-workers have shown that it is possible to efficiently immobilise montmorillonite onto cotton textile through their functionalization with organosilanes, thus increasing washing fastness[12].

Promising results of above-mentioned studies recorded over the last few years and lack of literatures on the field of polyamide textiles coated with synthetic zeolites using industrially-applicable procedures convinced us to research the viability of fabricating zeolite-modified fibre-forming polymers with changed water absorption properties. For the permanent surface coating of fibrous materials by zeolites with the aim of either improving some undesired properties or imparting advanced functionalities, numerous requirements must be taken into consideration such as appropriate zeolite structures and properties (type, pore size, adsorption capacity, etc.), textile composition and construction, pre-treatment of textiles, zeolite application procedure, as well as textiles after-treatment. Therefore, the main objective of our study was to select and characterise four different types of commercially-available zeolites for further surface treatment of polyamide knitted fabric using a novel dyeing-like exhaustion procedure in order to create (super)hydrophilic feature. As coating procedures can cause yellowness or colour changes indicating surface damage, and because visual appearance is also very important parameter, the optical properties were evaluated.

2.0. Experimental

2.1. Materials

Four different commercially-available zeolites were used in the forms of very fine powder, i.e. 4A, 13X, ZSM-5 and a mixture of those three zeolites in a ratio 60:30:10, which were industrially manufactured and supplied by the Silkem, Inc., Slovenia. Zeolite particles of average particle sizes of between 3 and 5 μm, and maximum particles size up to 10 μm, and typical chemical composition of 17-19 wt.% Na_2O, 28-30 wt.% Al_2O_3, 31-40 wt.% SiO_2, 18-22 wt.% H_2O were synthesised according to hydrothermal treatment of the aluminosilicate gel containing different organic structure directing agents.[7]

A light grey 100% polyamide (PA) knitted fabric was employed for all coating trials, and was industrially-manufactured by Julon d.d., Slovenia.

2.2. Coating procedure

Eight previously-optimised initial baths were composed of: i) 3% owf (of weight of fabric) of individual zeolite 4A, 13X, ZSM-5 or mixture of zeolites, and ii) 3% owf of individual zeolite 4A, 13X, ZSM-5 or mixture of zeolites together with suitable additives (add.), i.e. 0.3% owf of amphoteric levelling agent (Keriolon A2N, Bezema), 0.5 mL L^{-1} of pH-regulator (Meropan OFS, Bezema) and CH_3COOH (80%) for pH 5.5-6.5 adjustment. Polyamide knitted fabric was modified with an individual bath according to the novel industrially-acceptable exhaustion procedure at a temperature of 98°C for 60 min using a liquor-to-fabric weight ratio of 30:1 within a sealed stainless-steel treatment-pot, housed in a Labomat (W. Mathis). The treated samples were washed at a temperature of 40°C for 30 min in a Labomat (W. Mathis) using a liquor-to-fabric weight ratio of 50:1 in a solution of 4 g L^{-1} of standard reference detergent without optical brighteners. Then, the samples were rinsed in tap water, and dried at room temperature.

2.3. Characterization of zeolites

SEM analysis was carried-out by putting an individual zeolite powder on adhesive carbon tape, which was placed on a brass holder, and observed on a FE-SEM-ZEISS Gemini Supra 35 VP scanning electron microscope (Carl Zeiss NTS GmbH, Germany).

The infrared absorbance spectra of the zeolites were obtained using a spectrophotometer FTIR System Spectrum GX (Perkin Elmer) with a Golden Gate ATR attachment and a diamond crystal. The measurements were taken within the range of 4000-650 cm^{-1} wavenumber using 16 scans and a resolution of 4 cm^{-1}.

The BET specific surface area (S_{BET}) and porosity of the individual zeolites were determined by nitrogen adsorption-desorption isotherms at 77 K using the Tristar 3000 (Micromeritics) system according to standard ISO 9277. A detailed description of the nitrogen physiosorption measuring procedure is disclosed in.[7] Prior to these measurements, the samples were dried for 24 hours at a temperature of 200°C.

2.4. Fabric's surface characterisation

SEM was utilised for studying the surface morphologies of PA upgraded with zeolites. Approximately 1cm² of the fabric was attached on an adhesive carbon band on a brass holder on a ZeissGeminiSupra 35 VP Scanning Electron Microscope (Carl Zeiss NTS GmbH, Germany), and then the SEM images were taken. The infrared absorbance spectra of reference and the zeolites-coated PA were obtained using a spectrophotometer FTIR System Spectrum GX (Perkin Elmer) with a Golden Gate ATR attachment and a diamond crystal.

Table 1: BET specific surface area and porosity analyses

Sample	S_{BET} (m^2 g^{-1})	V_{mi} (cm^3 g^{-1})	V_{tot} (cm^3 g^{-1})
13X	629	0.290	0.310
4A*	3	-	0.003
ZSM-5	343	0.125	0.178
mixture	204	0.095	0.102

*sample should be dried at higher temperature by another apparatus

The measurements were taken within the range of 4000-650 cm^{-1} wavenumber using 32 scans.

The diffuse reflectance spectra profiles of samples in the 200-700 nm wavebands were recorded on a Lambda 900 UV-Vis-NIR spectrophotometer (Perkin Elmer) with and integrated sphere. CIE measurements of lightness (L^*) and chroma (C^*) of (un)coated samples were achieved within a spectral range of 400-700 nm wavelengths by a two-ray Spectraflash SF600 Plus spectrophotometer (Datacolor) equipped with an Ulbricht sphere and measuring geometry of d/8° under a standard illuminant D65 (LAV/Spec. Incl.).

The hydrophilicity of the untreated and zeolite-modified polyamide was evaluated with respect to the contact angle measurement using a sessile drop technique. An individual sample was placed on a horizontal table attached to a mechanical device on a Goniometer (DataphysicApparatus). A micro-drop with a volume of 3 μL MilliQ water was poured onto the fabric's surface. The drop was illuminated by white diffuse light and observed with tele-microscope. A clear image of the drop was directly transferred through a CCD-camera showing the drop profile. The contact angle was determined from the tangent to the drop at the triple phase contact line. For each sample, 5 repeated measurements were performed in order to obtain the mean value.

3.0. Results and Discussion

3.1. Analysis of Zeolites

SEM micrographs of zeolites were taken (Fig. 1), with the aim of studying the surface morphologies of four different zeolites for the subsequent surface modification of a selected fibre-forming polymer. Additionally, ATR-FTIR spectra of zeolites powder were recorded (Fig. 2) and gas physiosorption determined (Table 1).

SEM images in Fig. 1 show the structural/formational differences between the examined zeolites, namely cubic crystals of zeolite A (a), octahedral crystals of zeolite X (b) and pentasil building blocks of zeolite ZSM-5 (c), which were all present in the mixture (d) as expected from the producer's information. Furthermore, a high amount of sodalite was noticed as well as the presence of impurities, probably due to the industrial synthesis procedure using real industrial products (not laboratory grade chemicals) with lower purities as sources for zeolites synthesis. It was also noted from Fig. 1 that the examined zeolites were different sizes ranging from approximately 0.5 up to 6 μm.

It can be seen from the FTIR spectra of zeolites in Fig. 2 that intensive bands appeared in the area of low frequency at a wavenumber of ca. 970 cm^{-1} (4A, 13X and ZSM-5) and at 1067 (mixture) and corresponded to the stretching vibrations of Si–O(Si) and Al–O(H) functional groups, respectively[13]. In addition, typical bands can be noted within the range of 680-800 cm^{-1}, regarding the type of zeolite, as a result of the 'ring' vibrations of internal oxygen bridges Si–O–Al and Si–O–Si; and at a wavenumber of ca. 3340 cm^{-1} due to the vibrations of H–O band of the absorbed moisture.

Fig. 1: The SEM images of selected zeolites **(a)** 4A **(b)** 13X **(c)** ZSM-5 **(d)** mixture

Table 1 gathered typical micropore volume and total pore volume for zeolites ZSM-5 and 13X. On the other hand, zeolite 4A had the lowest values of BET surface as well as total pore volume, probably due to the filled pores and should thus, be dried at higher temperatures. Moreover, for the mixture of zeolites the three measured parameters were rather low as expected because the ratio of 4A, 13X and ZSM-5 was 60:30:10.

3.2. Morphological Molecular-Chemical Analyses of Zeolites-Coated PA Fabrics

In order to visually-establish the amount/equality of the four zeolites applied on the surface of PA according to a novel dyeing-like procedure with or without the presence of selected additives, the surface morphologies were studied by SEM (Fig. 3) and the molecular-chemical changes by FTIR (Fig. 4).

The SEM images in Fig. 3 demonstrated different coating morphologies regarding the zeolite types, i.e. 4A, 13X, ZSM-5, and the zeolites mixture, as well as the employed baths compositions, i.e. aqueous dispersions of individual zeolites (1st row), and slightly acidic aqueous dispersions of individual zeolites together with proper additives - add. (2nd row). In the cases of aqueous dispersions of zeolites 4A and 13X, which were very alkaline with pH 11.5, exceedingly low content of micro-particles could be observed over complete fibers' surfaces as compared to the samples treated with the same zeolites in combination with additives upon the acidic conditions (pH 5.5-6.5). Under acidic conditions, the amino groups at the ends of the polyamide chains are protonated, and under highly alkaline medium the end-carboxylic groups become negatively charged.[2,3] This amphoteric character of fibers could be involved within the interactions with zeolite particles. The obtained SEM results were in good agreement with the FTIR and DRS analyses additionally presented in Figs. 4 and 5.

In Fig. 4, typical peak positions for polyamide are depicted including amide I (C=O stretching vibrations) at ~1631 cm^{-1} and amide II (N–H bending and C–H stretching vibrations) at 1531 cm^{-1} as also fully interpreted[1,6].

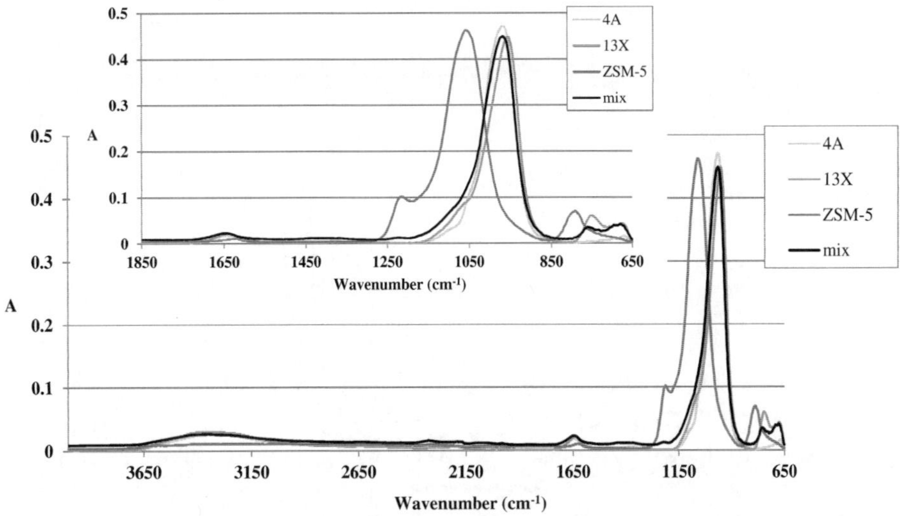

Fig. 2: ATR-FTIR spectra of used zeolites

Fig. 3: SEM images of PA fabrics coated with 3% owf of: **(a)** zeolite 4A **(b)** zeolite 13X **(c)** zeolite ZSM-5 **(d)** zeolite mix **(e)** zeolite 4A + add **(f)** zeolite 13X + add **(g)** zeolite ZSM-5 + add. **(h)** zeolite mix. + add.

In addition, some characteristic absorption bands were evidently recognised for the zeolite modified PA, i.e. the peaks around 1000 or 1060 cm^{-1} (regarding the zeolite type) corresponded to stretching vibrations and at 799 cm^{-1} to the bending Si–O vibrations.[4,6] Meanwhile, some peaks became lower, such as peaks at 1631 cm^{-1}, 1531 cm^{-1} and 3298 cm^{-1} due to the reduction of carboxylic and amino end groups of polyamide fibres or intensified, such as peaks at 1195-1270 cm^{-1} on account of the presence of bending vibrations of Si–O–CO–CH_2–.

3.3. Diffuse Reflectance Spectra Profile Determination and CIE Measurement

The DRS profiles of the polyamide within the UV and Vis regions (wavelengths between 250 and 700 nm) before and after the zeolites' modifications are compared in Fig. 5.

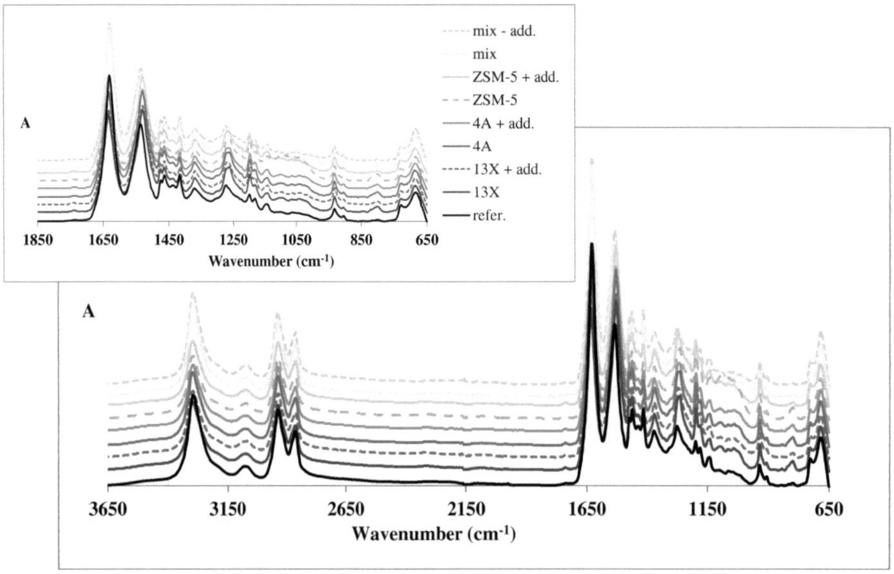

Fig. 4: FTIR spectra of (refer.) and coated PA using aqueous dispersions of different zeolites and zeolites together with additives (add.).

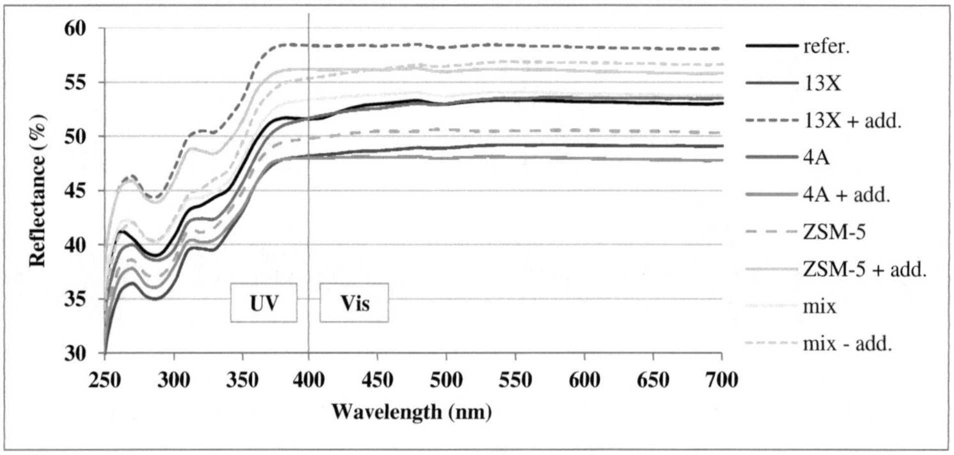

Fig. 5: Diffuse reflectance spectra of untreated (refer.) and coated PA using aqueous dispersions of different zeolites and zeolites together with additives (add.).

Moreover, based on the presumption that zeolites are white pigments and thus applied on the material's surface producing a white colour or that coating can cause yellowness/colour changes indicating surface damages, the influences of the coatings of different zeolites on the samples' visual changes were given by CIE lightness (L^*) and chroma (C^*) determination (Fig. 6).

Fig. 5 shows a reflectance curve of bright grey untreated fabric (ref.) with one typical absorption peak (reflectance minimum) at wavelength of 290 nm, which are contributed to the optical profile of the polyamide. After the zeolites' applications, the optical properties were changed in both, UV and visible regions regarding the zeolite type and treatment bath composition.

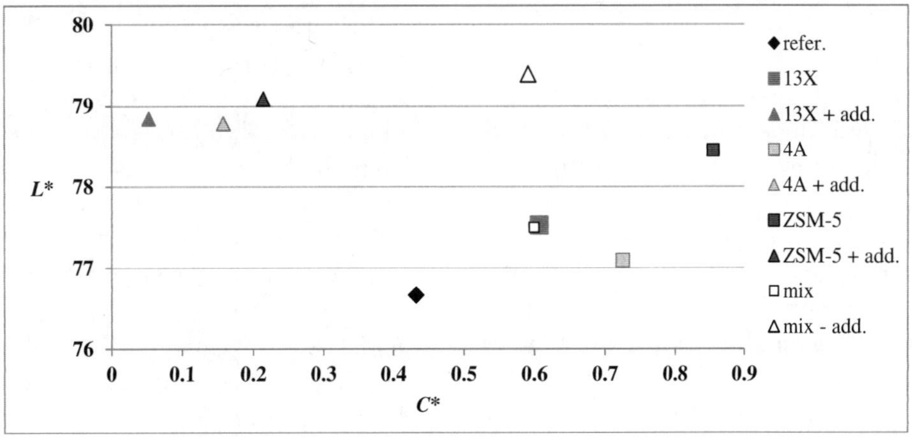

Fig. 6: CIE lightness (L^*) versus chroma (C^*) of untreated (refer.) and coated PA using aqueous dispersions of different zeolites and zeolites together with additives (add.).

Table 2: Water contact angles (CA) of untreated (refer.) and coated PA using aqueous dispersions of different zeolites and zeolites together with additives (add.)

Sample	CA [°]	Drop images
untreated	107.3	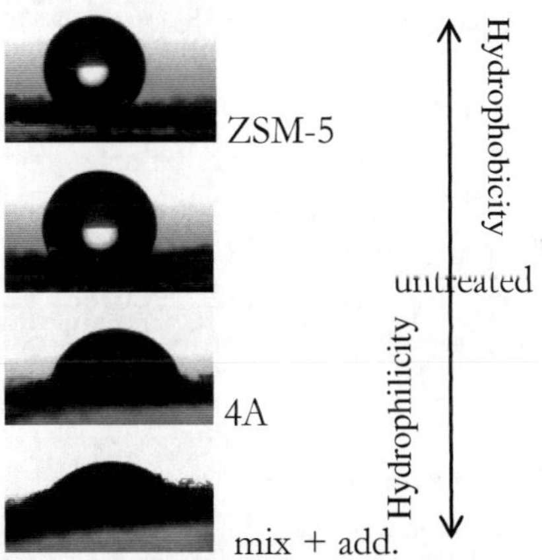
4A	54.5	
4A + add.	8.6	
13X	127.1	
13X + add.	18.0	
ZSM-5	132.8	
ZSM-5 + add.	20.3	
mix	78.6	
mix + add.	16.3	

Also, an extra absorption peak occurred within the UV-A region at 330 nm for all the coated samples. Moreover, the reflectance curves of fabrics treated only with zeolites 13X, 4A or ZSM-5 decreased most within the spectrum's UV region, implying stronger absorption intensity in comparison with the reference, as well as those samples treated with the combination of individual zeolites and additives.

The results in Fig. 6 revealed visually perceivable changes in lightness (and somewhat less in chroma) between the reference and zeolite-coated samples as well as between the baths' compositions. As expected, the samples treated with slightly acidic (pH 5.5-6.5) aqueous dispersion of zeolites in combination with appropriate additives were brighter (L^* values were higher) and less saturated (C^* values were lower) compared to the reference revealing higher amounts of zeolites on the surfaces of the fibres, as could also be perceived from the SEM. The most chromatic sample was treated with 3% owf of ZSM-5 aqueous dispersion.

3.4. Water contact angle determination

The changes in the water absorption properties of the zeolite-coated polyamides compared with the untreated ones were determined in terms of the water droplet (3 μL) contact angle by means of goniometry and the gained results are disclosed in Table 2.

From Table 2, it could be summarised that water absorption properties varied significantly depending on the zeolite type as well as the baths' compositions. Industrially-manufactured non-coated knitted polyamide fabric showed an average contact angle of about 107.3°. Applications of zeolites ZSM-5 and 13X aqueous dispersions enlarged the contact angles by up to 132.8° and 127.1°, respectively, and thus, PA became more hydrophobic, whilest the zeolites mixture's employment reduced the CA to 78.6° and 4A to 54.5° rendering the PA fabric more hydrophilic in comparison to the reference sample.

On the other hand, the application of 3% owf of the zeolite acidic bath (irrespectively of zeolite type used) together with selected auxiliaries imposingly reduced the water contact angles to about 8.6-20.3°. In those cases, the droplets absorbed completely by the modified fabrics after ca. 10 s. Carran et al. reported that the acidic condition can caused the dealumination of the zeolites, resulting in the breakdown of the zeolite structures and thus generating a greater amount of –OH containing species of which some can be bound to the PA surface, and the other caused enhanced water absorption abilities of the fibres.[7]

4.0. Conclusions

- From the SEM, FTIR and gas physiosorption analyses of used industrially-synthesised zeolites, it could be observed different morphologies and chemical characteristics on a molecular level, which was important for further fabric surface functionalization.
- Surface morphology observations and identifications of specific molecular vibrations of the zeolite-coated polyamide knitted fabrics proved the suitability of the employed exhaustion procedure regarding the type of zeolite used as well as the compositions of the treatment baths.
- The optical properties of the zeolite-coated fabrics were changed in both the UV and visible regions, i.e. zeolite-treated samples became lighter.
- Water absorption properties varied significantly depending on the zeolite type as well as the baths' compositions.
- Applications of ZSM-5 and 13X alkaline aqueous dispersions enlarged the contact angles up to 132.8° and 127.1°, respectively.
- Zeolite acidic bath (irrespective of the zeolite type used) together with selected auxiliaries imposingly reduced the water contact angles to about 8.6-20.3° - fabricated zeolite-PA fabrics became exceedingly hydrophilic compared to the non-coated sample.

5.0. Acknowledgements

The research leading to these results has received funding from the Slovenian Research Agency (ARRS) under applied project no. L2-6776, and a programme group for Textile Chemistry P2-0118.

6.0. References

1. Cloud, R.M.; Song, G.; Functional finishes to improve the comfort and protection of apparel, in Book: Advances in the dyeing and finishing of technical textiles, M.L. Gulrajani, ed., Wodhead Publishing Limited, Cambridge, UK, 2013.

2. Christie, R.M.; Colour Chemistry, 2nd edition, Royal Society of Chemistry, Cambridge, UK, 2015.
3. Rouette, H.K.; Encyclopedia of textile finishing, Springer, Aachen, Germany, 2000.
4. Kim, H.R.; Seo, H.Y.; Textile Res. J., vol. 83(11), p. 1181.
5. Fakin, D.; Stana Kleinschek, K.; Ojstršek, A.; Acta Physica Polonica A, 2015, vol. 127(4), p. 943.
6. Liu, S.; Cao, X.; Li, L.; Li, C.; Ji, Y.; Xiao, F.S.; Colloids and Surfaces A: Physicochemical and Engineering Aspects, 2008, vol. 318(1-3), p. 269.
7. Fakin, T.; Ristić, A.; Mavrodinova, V.; Zabukovec Logar, N.; Microporous and Mesoporous Materials, 2015, vol. 213, p. 108.
8. Bastani, D.; Esmaeili, N.; Asadollahi, M.; J. Ind. Eng. Chem., 2013, vol. 19(2), p. 375.
9. Grancarić, A.M.; Tarbuk, A.; Kovaček, I.; Chem. Ind. Chem. Eng. Quarterly, 2009, vol. 15(4), p. 203.
10. Grancarić, A.M.; Tarbuk, A.; McCall, D.; Polimeri, 2007, vol. 28(4), p. 219.
11. Carran, R.S.; Ghosh, A.; Dyer, J.M.; Appl. Surf. Sci., 2013, vol. 287, p. 467.
12. PMonteiro, A.; Jarrais, B.; Rocha, I.M.; Pereira, C.; Pereira, M.F.R.; Freire, C.; Appl. Clay Sci., 2014, vol. 101, p. 304.
13. Mozgava, W.; J. Mol. Struc., 2001, vol. 596(1-3), p. 129.

Modeling of the Input Parameters and Investigation of the Surface Residual Stresses and Deformations in Deep Rolling

N. Lyubenova*, L. Thuillier, D. Baehre
Saarland University,
Institute of Production Engineering, Saarbruecken, Germany
*Email: nataliya.lyubenova@uni-saarland.de

Abstract

Deep Rolling (DR) is a widely used process that induces compressive (=negative) residual stresses on the surface and in the depth of materials. It is well known that the targeted application of compressive residual stresses in highly stressed components is often used to improve their fatigue strength. As the tensile (=positive) loads promote the crack formations, the compressive residual stresses retard the surface cracking and prevent the initiation of new cracks, thus increasing the components' endurance.

A suitable tool to model complex processes, like DR, is the Finite Element (FE) Modeling. It offers a good opportunity to vary and investigate numerous input and output parameters of the process. The presented investigations use the Explicit module of the FE-Code ABAQUS 6.14 to model the DR process on a plane specimen manufactured from AISI 4140 steel. The applied normal force and the percentage of overlapping are varied and their influence on the induced surface residual stresses as well as the resulting surface deformations are commented on.

1.0. Introduction

Deep Rolling (DR) is an established process that induces compressive surface and sub-surface residual stresses[1]. It improves also the quality of the treated surface and creates a cold working. Its main principle is shown in Fig. 1(a) and the typical depth distribution of the induced residual stresses in Fig. 1(b).

Although well-known, DR is still investigated due to the large amount of existing input parameters which lead to great differences in the results[2]. Some of them, like the applied normal force or pressure, the number of overturns and the percentage of overlapping, directly influence the quality of the surface as well as the distribution of the induced residual stresses.

Using the FE-Modeling, numerous input parameters as well as geometries can be varied. Baecker et al. used a FE-Modell to vary the applied DR pressure on a plane geometry and found out that the increasing of the pressure leads to a higher maximum of the induced compressive residual stresses below the surface[3]. FE-Modeling of the DR process on an axis-symmetrical component was introduced by Trauth et al.[4]. They compared the modelled depth profile of residual stresses with experimental data and reported a good agreement between both. Though several attempts to model the DR process are made, all of them focus on the depth distribution of the induced residual stresses[3–5]. If considering that the smooth surface distribution of residual stresses is also important for the surface integrity, the investigation and parameter variation to optimize the residual stress surface distribution is also required. Therefore, this paper shows the surface residual stress distribution when varying the applied force or percentage of overlapping. Another problem also raised here is the possible deformation (bending) due to the DR. Klocke et al. reported for possible bending of the turbine blades treated with DR, but his statement was not quantified with an investigation[6]. For this reason, a study of a possible bending due to the DR is also shown in the presented paper.

2.0. Finite Element modeling of Deep Rolling

2.1. Geometries

The DR process was modelled using a three dimensional work piece and a DR sphere, see Fig. 2. Due to the complexity of the process and the anisotropy of the residual stresses induced in rolling and transverse direction, a two dimensional representation was not employed[5].

2.2. Material description

To describe the characteristic material behaviour, the uniaxial tensile-compressive test is commonly used[7]. Huang shows this model with a realistic non-linear and a bi-linear approximation of the material behaviour[8]. Fig. 3 describes the four main regions the realistic stress-strain curve is divided into. Due to the Bauschinger effect, the yielding in the unloading phase B-C starts earlier than in the loading phase 0-A, depending on the formerly experienced plastic strain during the loading[9].

Different linear approximating material behaviours are also represented in Fig. 3. The first representation that considered the Bauschinger-Effect is closest to the real material behaviour. Still, this option is difficult to apply as the Bauschinger coefficient has to be experimentally derived. The second option - elastic-plastic representation with bi-linear kinematic hardening - offers a close approximation to the real material behaviour and will be therefore used in the presented investigations[10]. The last variant - linear elastic, plastic with isotropic hardening - does not consider the Bauschinger-Effect and its implementation in the FE-model can lead to an overestimation of the calculated residual stresses.

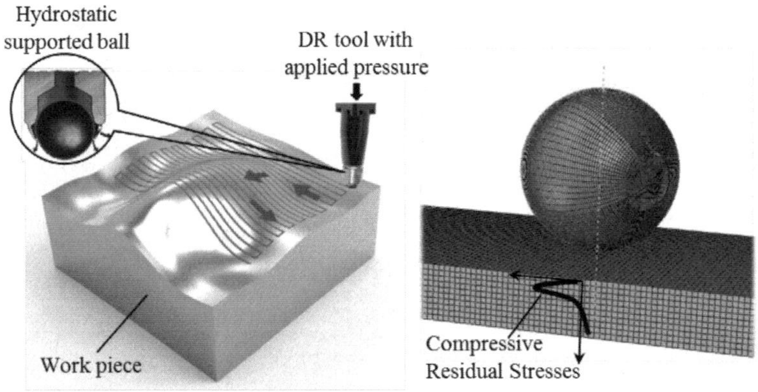

Fig. 1: **(a)** DR principle **(b)** induced residual stresses

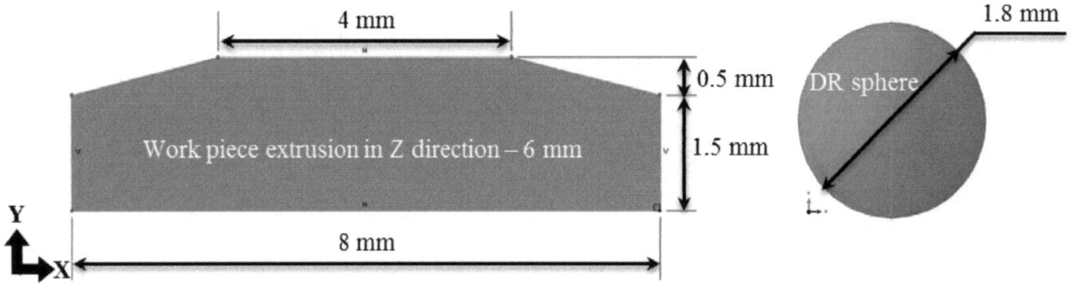

Fig. 2: Geometry of the work piece and DR sphere

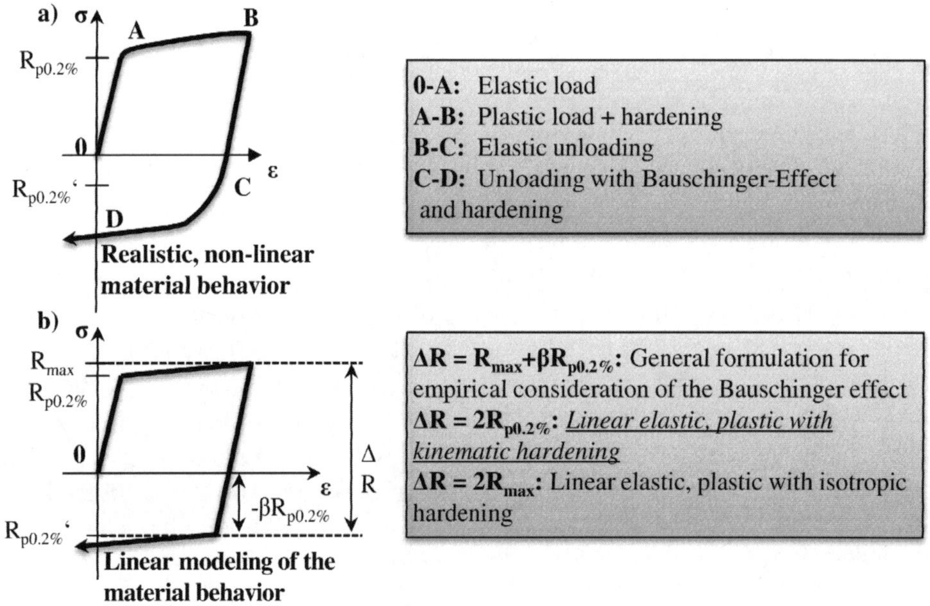

Fig. 3: Material description [10]

Table 1: Material data

DIN EN 10002 uniaxial tensile test data for AISI 4140	
Young's modulus	211 000 MPa
Yield strength	973 MPa
Ultimate strength	1 073 MPa
Fracture strain	4.72%
Poisson's ratio	0.28

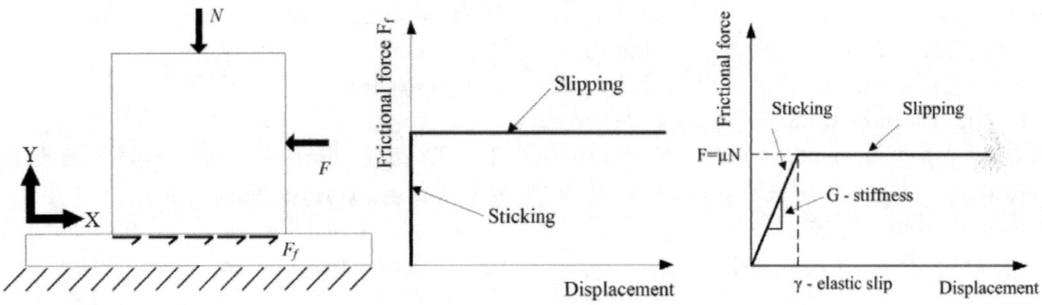

Fig. 4: (a) Basic Coulomb friction model **(b)** Coulomb friction ideal curve **(c)** Friction curve with penalty formulation[11].

For the material of the work piece, a high strength steel type AISI 4140 was employed. The material data extracted from uniaxial tensile test is presented in Table 1. The DR sphere was modeled as a rigid body.

2.3. Meshing

In the presented model numerous mesh densities as well as different elements were varied to ensure proper and not too extensive calculation times as well as to minimize the meshing as a source of errors. Finally, the work piece was meshed using hexahedral C3D8R elements. The smallest element size was 0.025 mm and the largest one – 0.1 mm. To ensure a smooth decreasing of the mesh density out of the area of interest, a progressive meshing strategy was applied.

2.4. Contact and friction formulation

In the DR process at least two components are in contact with each other: the work piece and the DR tool. The proper modeling of the contact between both components is important as this can lead to different surface deformations and near-surface stress distributions. ABAQUS offers different possibilities to model a contact; one of them is the "contact pair" option. Here, the two surfaces which will be in contact have to be assigned and defined as master (DR sphere) and slave (work piece) surfaces. After the definition of the contact pairs the friction between them has to be formulated.

In ABAQUS the used friction model is based on the Coulomb friction model, as described in Figs. 4 (a) and (b).

Equation 1.1 shows the dependence of the applied normal force **N** on the resulting frictional force $\mathbf{F_f}$, whereas the applied tangential force **F** is equal to $\mathbf{F_f}$. The friction coefficient **μ** is a scalar that describes the friction ratio between two surfaces. In this paper it was fixed to 0.1.

$$F_f = \mu.N \qquad (1.1)$$

The Coulomb friction model in ABAQUS is extended to fit different simulation cases. One of those extensions is the penalty formulation, see Figure 4 c). It describes the friction stiffness that allows a relative motion, i.e. the elastic slip **γ**, of the contacted surfaces when they should remain sticking. The elastic slip is an important parameter as it affects the friction behaviour before the slipping phase. The elastic slip is defined as the distance where the slipping phase begins. The stiffness of the inclined line **G** is described in Equation 1.2.

$$G = \mu.N.\gamma^{-1} \qquad (1.2)$$

Under realistic conditions the elastic slip can be attributed to the existing elastic displacements in the surface roughness. In ABAQUS the elastic slip can be defined as a fracture of the element length or as an absolute distance. In the presented investigations the first option was applied and the elastic slip was fixed to 0.005%. With regard to the frictional directionality, the isotropic option was used.

2.5. Modeling steps and boundary conditions

The presented DR modeling consists of 7 steps:

Initial step: the work piece was fixed in all directions and the DR sphere was allowed to move in vertical Y direction.

Contact step: a contact was assigned between the work piece and the DR sphere as applying a normal force of 5 N in -Y direction.

Apply force step: the contact force was deactivated and the application of the DR force in -Y direction took place. Depending on the different investigations, the DR force was varied from 25 to 100 N.

Movement 1: the DR sphere moved in Z direction. Its rotation was free and a translational velocity of v = 1 $mm.s^{-1}$ was defined.

Movement 2: the DR sphere moved in X direction. Its rotation was free and a translational velocity of v = 1 $mm.s^{-1}$ was defined.

Movement 3: the DR sphere moved in -Z direction. Its rotation was free and a translational velocity of v = 1 $mm.s^{-1}$ was defined.

Release force step: the DR force was released.

3.0. Results

3.1. Surface distribution of residual stresses – DR normal force variation

The first result shows the surface distribution of stresses when the DR force is varied, see Fig. 5. Here, the application of a higher force does not lead to higher compressive residual stresses. The higher DR force still induces a brighter area with residual stress.

3.2. Surface distribution of residual stresses – variation of the overlapping

In Fig. 6, the surface stress distribution after variation of the overlapping is plotted. Here, the applied DR force was 100 N. The results show that an overlapping from 25 to 75% leads to relatively smooth surface residual stresses. Also the different percentage of the overlapping does not change the maximum of the surface residual stresses.

3.3. Surface deformations after releasing of the work piece boundary conditions

The aim of the presented test is to investigate a possible bending of the work piece due to the DR process.

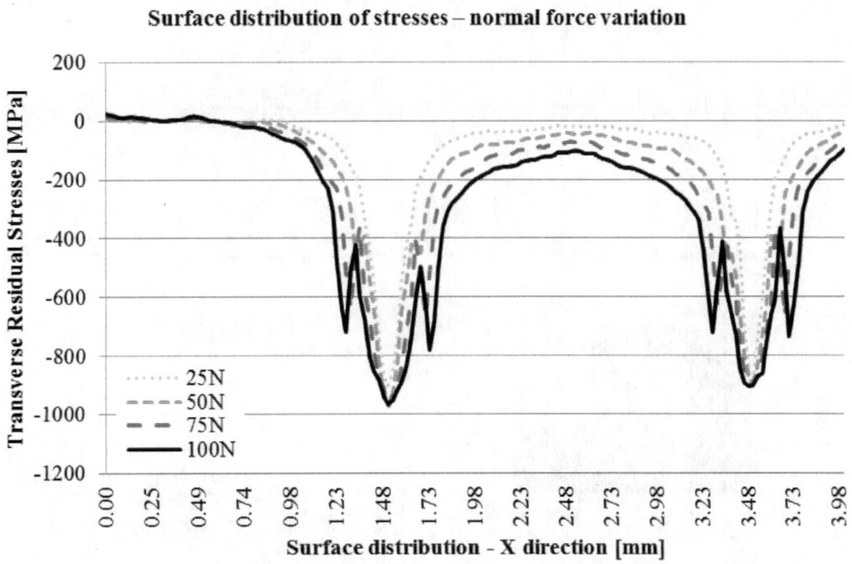

Fig. 5: Surface distribution of stresses – variation of the normal force

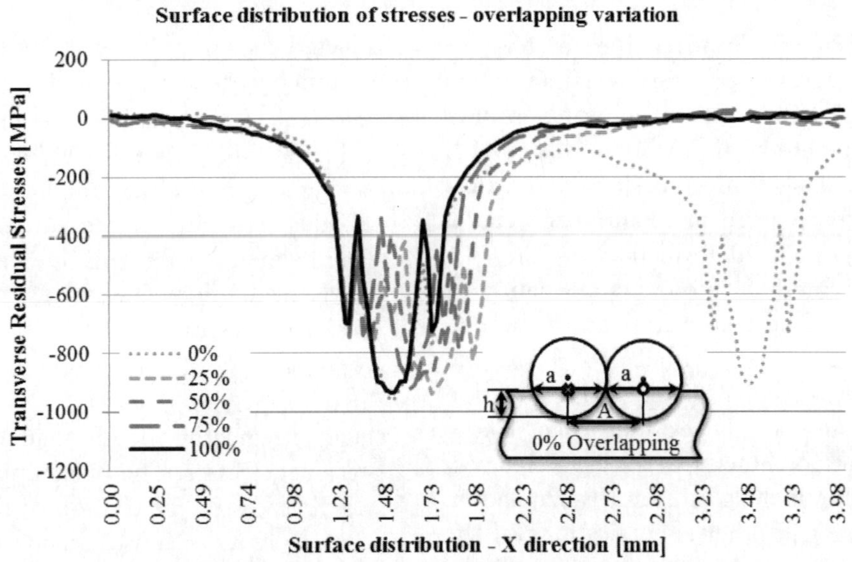

Fig. 6: Surface distribution of stresses - overlapping variation

After DR the work piece fixation was deactivated and then the resulting deformations were investigated. In Fig. 7 the displacement in Y direction over the surface distribution in X direction is plotted. Here, it is visible that along with the deformations due to the DR sphere penetration no additional bending can be observed.

4.0. Discussion

In all of the presented results some sharp peaks of the residual stresses and the displacements are visible. This can be attributed to the DR process itself but they can be also artefacts from the mesh quality.

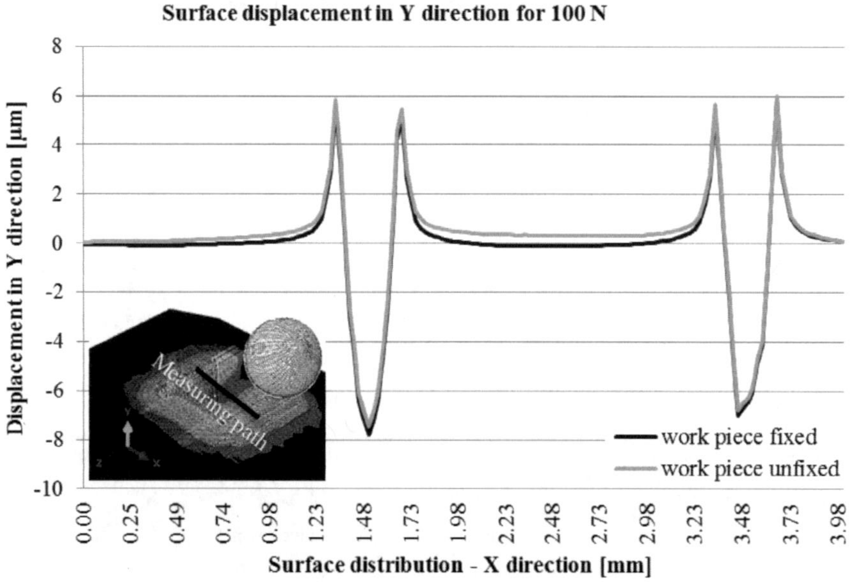

Fig. 7: Surface displacement in Y direction

One possibility to improve the mesh is to increase its density, but this will lead to an increased calculation time. Therefore, another option can be used – the Arbitrary Lagrangian Eulerian (ALE) adaptive meshing. It offers the possibility to maintain a high-quality mesh, even when large deformation occurs, by allowing the mesh to move independently of the material[12]. Another critical part of the FE-Model is the contact formulation. Here, the DR sphere was fixed in the normal Y direction after application of the normal force. Under realistic conditions the DR tool has a free play that should be included in the FE-simulation. Due to these and other simplifications of the FE-Model as well as the insufficient corresponding literature, it is highly recommended to conduct experimental data to be compared with the presented model.

5.0. Conclusion

In this paper the surface distribution of residual stresses after DR on a plane geometry was shown as well as the deformations due to the process. Three investigations were conducted: a normal force variation, an overlapping variation and a deformation study.

- Normal force variation: the application of a higher DR normal force does not lead to a higher maximum of the compressive residual stresses on the surface. Still, the employment of a higher DR normal force leads to a brighter area with compressive residual stresses.
- Variation of the overlapping: it is clear that no overlapping (Fig. 6, 0 %) leads to high irregularity of the induced surface residual stresses. The overlapping with just 25 % leads to a generation of the brightest area with residual stress and offers also a faster DR treatment of a larger zone compared with 75 % overlapping.
- Deformations due to the DR: here, two observations can be made. First, the DR sphere creates a deformation of about 14 µm that is reproducible if the applied force remains the same, see the surface distribution in X direction – 1.48 mm and 3.48 mm. Second, the maximal deformation after the release of the work piece fixture is about 0.5 µm at 2.48 mm in

X direction, which is negligible compared to the deformations due to the indentation of the DR sphere.

6.0. References

1. Altenberger, I.; Deep Rolling – The past, the present and the future, Conference Proceedings: ICSP-9, 2005.
2. Schulze, V.; Procedures of mechanical surface treatments, Modern mechanical surface treatment. Wiley-VCH, 2006.
3. Baecker, V.; Klocke, F.; Wegner, H.; Timmer, A.; Grzhibovskis, R.; Rjasanow, S.; Mater. Sci. Eng., 2010, vol. 10, p. 012134.
4. Trauth, D.; Klocke, F.; Mattfeld, P.; Klink. A.; Procedia, CIRP 9, 2013.
5. Manouchehrifar, A.; Alasvandf, K.; Recent Researches in Engineering Mechanics, Urban & Naval Transportation and Tourism, 2009.
6. Klocke, F.; Baecker, V.; Wegner, H.; Zimmermann, M.; Proceedings of the Institution of Mechanical Engineers, Part B: J. Eng. Manuf., 2011, vol. 225(1), 2011.
7. Macherauch, E.; Praktikum in Werkstoffkunde, Vieweg Verlag, 1987.
8. Huang, X.P.; The J. of Strain Analy. Eng. Des., 2005.
9. Bauschinger, J.; Mittheilungen aus dem Mechanisch-technischen Laboratorium. K. TH München 1886, vol. 13.
10. Brünnet, H.; Dissertation, Schriftenreihe Fertigung und Konstruktion Band 50, 2014.
11. ABAQUS 6.14 analysis user's guide, 'Chapter 37.1.5 - Using the basic Coulomb friction model,' 2014.
12. ABAQUS 6.13 analysis user's guide, 'Chapter 12.2.1 ALE adaptive meshing: overview,' 2013.

Towards More Eco-Friendly Fabrication Approaches for Superhydrophobic and Superoleophobic Coatings

A. Milionis*,
Mechanical and Aerospace Engineering,
University of Virginia, Charlottesville (VA), United States.

I. S. Bayer,
Mechanical and Aerospace Engineering,
University of Virginia, Charlottesville (VA), United States.

Smart Materials, Istituto Italiano di Tecnologia, Genoa, Italy.

E. Loth
Mechanical and Aerospace Engineering,
University of Virginia, Charlottesville (VA), United States.
*Email: am2vy@virginia.edu

Abstract

In this work we present two spray-based fabrication routes to achieve superhydrophobic and superoleophobic polymer/particulate coatings. The first two-step approach includes the use of a commercial bioplastic and hydrophobic nano-silica, and subsequent chemical functionalization with a fluoroacrylic copolymer. The second single-step approach is performed by spraying a waterborne dispersion of hydrophilic nano-silica and acrylic fluoropolymer. The acrylic fluopolymer used here has C-6 chemistry and does not degrade to toxic byproducts. Both surfaces exhibit static water and oil contact angles greater than 150° and low roll off angles.

1.0. Introduction

Technologies related to liquid-repellent materials have drawn significant attention and research efforts during the past years due to their vast number of commercial, industrial and military potential applications. These efforts have led to a large number of publications as well as new commercial products[1]. Generally, surfaces are referred to as superhydrophobic if they display water contact angles (WCA) > 150° and very low droplet roll off angles, and if they have the same performance with oils, they are referred to as superoleophobic.

Although all these studies are successful in terms of application, very few of them use materials that can be classified as environmentally friendly or follow green fabrication processes. Only in the last few years such approaches have become available for liquid-repellent materials and coatings. Other biodegradable materials that have been used for superhydrophobic surfaces include cellulose, poly(lactic acid), hydroxycaproic acid, magnesium alloys and poly(3-caprolactone) and their fabrication techniques include plasma processing, chemical vapor deposition, in situ nanoparticle growth, electrospinning, phase separation, polymerization, spay and dip-coating using various solvents[2–5]. Again, most of these techniques require multiple fabrication steps, are difficult to scale-up, use flammable or hazardous solvents and the majority of them are surface functionalizations rather than polymer nanocomposite coatings.

On the other hand, superoleophobicity is even more challenging to achieve with sustainable materials since there is no natural surface that can completely repel oil (excluding underwater systems). In fact, only underwater oil repellency has been reported by natural surfaces like fish-scales or the lower part of the lotus leaf. Re-entrant surface curvature, chemical composition and nanoscale texture are crucial parameters for designing artificial superoleophobic surfaces.

Repelling oils traditionally requires the use of fluorinated materials. This imposes serious environmental concerns as recent studies have shown that long-chain perfluorinated compounds are difficult to degrade and tend to bioaccumulate[6]. Furthermore, even if these substances degrade after a prolonged period, some of their byproducts are known to be toxic, such as perfluorooctanoic acid[7] and perfluorooctanesulfonate[8]. However, apart from the long-chain perfluorinated compounds, there are also other important concerns in many aforementioned techniques. Many of them require the use of toxic and flammable solvents or highly corrosive acids that can impose additional human health or fire safety issues. Alternative fabrication routes that are more sustainable and reduce the risks related to the safety of the materials or procedures required for obtaining oil-repellent surfaces are desirable instead. Such strategies involve the use of non-hazardous or flammable solvents, use of short-chain fluorinated compounds (C6 or less) or alternative materials that do not contain fluorine atoms.

Herein, we present two simple, one-step spray coating processes to fabricate superoleophobic polymer nanocomposites comprising fumed silica nanoparticles, Materi-Bi® starch-based bioplastic and Capstone® fluoroacrylic copolymer (C-6 chemistry). The wetting properties of the composites were controlled by varying the ratio of the polymer to nanoparticles and superhydrophobicity/superoleophobicity was achieved for specific constituent ratios. In this work only environmentally approved materials were used,[5] while in the second fabrication approach water was used as processing solvent.[9] It is anticipated that such coatings can be incorporated in the development of liquid repellent green materials for packaging and filtering along with many other commercial uses.

2.0. Methods

2.1. Bioplastic-based nanocomposite preparation procedure

For the starched-based polymer formulation Aerosil® R812 (Evonik Industries, Germany) nanoparticles and Mater-Bi® (Novamont, Italy) were dispersed in chloroform in 1 wt. % concentrations using ultrasonic processing for 30 seconds. The bioplastic to nanoparticle weight ratio was 2:3. After ultrasonic treatment a clear solution was obtained. The coatings were prepared by spray-coating the nanosuspensions. For the spray-coating, a VL double action, internal mix, siphon feed airbrush was used (Paasche, USA). After the spraying process, the bioplastic-based nanocomposites were cured at 130° C for 30 min. In order to render the coatings superoleophobic a dispersion of 0.2 wt% Casptone® ST-100 was prepared in acetone and was subsequently sprayed as a top coating. The extraction of the fluoroacrylic copolymer from its original aqueous dispersion is described elsewhere.[10]

2.2. Water-based nanocomposite preparation procedure

For the water-based dispersion 0.4 g of Aerosil® 200 nanoparticles were dispersed in 42 ml of deionized water using ultrasonic processing (Sper Scientific: 100004) for 8 min. Subsequently 3 ml of Capstone® ST-110 (Dupont, USA) (note that this is a different version of the same product) were added and the mixture was stirred for 1 minute. The nanosuspensions were sprayed on hot aluminum substrates with the same spray apparatus in 4 steps of 5 s each, with an interval of 30 s between each cycle in order to ensure that there is sufficient time for the water to evaporate. During the spraying process, the substrates were placed on a hotplate maintaining at a constant temperature of 300°C. This high temperature was necessary for melting the polymer and obtaining a more uniform coating, but also for avoiding dewetting effects due to slow evaporation of water upon spray.

2.3. Characterization

Water and oil contact angles (WCAs, OCAs) and droplet roll-off angles (ROAs) on the surfaces were measured by a video based optical contact angle measuring instrument. Ten microliter of deionized water and mineral oil (light (CAS 8042-47-5) from Fisher Scientific) droplets were gently placed on the surfaces. Measurements were conducted on three different locations and averaged for each sample. In order to measure ROAs, the substrates were tilted until the droplets started to roll off the surfaces. The substrate angle at the onset of droplet roll-away was recorded. The morphology of the textured surfaces was characterized by scanning electron microscopy (SEM; Model: FEI Quanta 650, USA). To eliminate charging effects, the samples were coated with a very thin (~20 nm) layer of Au/Pd by sputtering.

3.0. Results and discussion

After the spraying process a dual micro/nano rough surface topology was obtained for both sprayed nanocomposite formulations. Fig. 1 shows SEM images from the surface of (a) bioplastic/hydrophobic nano-silica coating, surface treated with Capstone ST-100 and (b) Capstone ST-110 with hydrophilic nano-silica. The thickness of the coatings was measured approximately 5 μm. As it can be clearly seen, multi-scale rough surface topologies were obtained due to the incorporation of the nano-silica particles. The surfaces exhibit remarkable anti-wetting characteristics. In particular the water and oil static contact angles are above 150° for both nanocomposites.

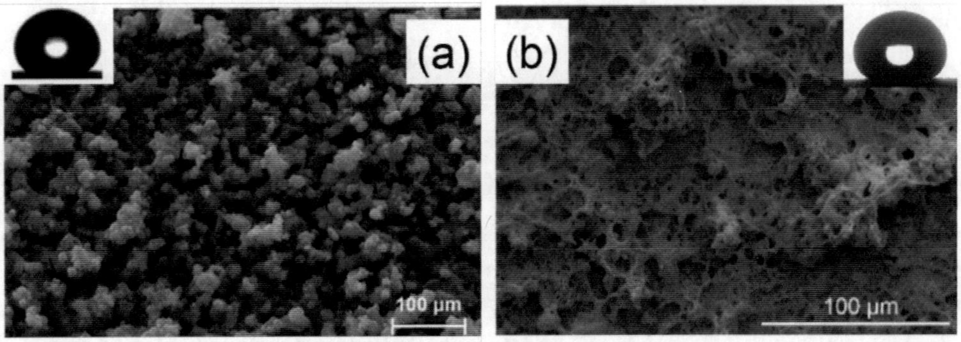

Fig. 1: SEM images of the surface topography for: (a) bioplastic/hydrophobic nano-silica coating, surface treated with Capstone ST-100, (b) Capstone ST-110 with hydrophilic nano-silica. Insets: oil droplets placed on the surfaces

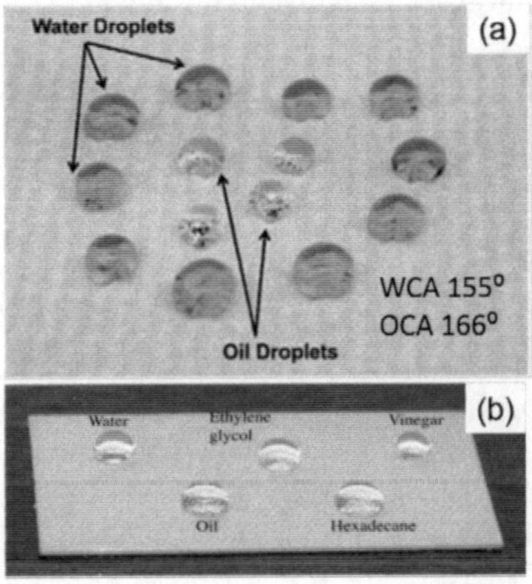

liquid	Contact angle (°)	Roll off angle (°)
water	159	2
oil	159	3
ethylene glycol	162	3
hexadecane	146	8
vinegar	157	5
coffee	159	4
milk	149	7
hot sauce	143	6
orange juice	159	4

Fig. 2: Photographs of different liquid droplets on: (a) bioplastic/hydrophobic nano-silica coating, surface treated with Capstone ST-100 on a cotton fabric, (b) Capstone ST-110 with hydrophilic nano-silica on aluminum. Right: Table with CAs and ROAs for various liquids on the coating.

These values were measured on aluminum substrates but similar values were obtained also in the case of other various substrates. We should mention here that the choice of Mater-Bi (for the first approach) was also because it is hydrophobic on top of its biodegradable properties. The WCA of a pure Mater-Bi coating without particles was measured approximately 80°. In terms of ROA the bioplastic-based nanocomposites exhibit values of 17° for water and 15° for oil (static WCA 155° and OCA 166°), while in the case of the coating sprayed with the waterborne dispersion, these values are much lower, 2° and 3° respectively (static WCA 159° and OCA 159°). This can be attributed to the difference in the surface topography as shown from the Fig. 1 where the bioplastic-based nanocomposite seems to exhibit rougher surface characteristics, but also due to the hydrophilic nature of the biopolymer.

The bioplastic-based nanocomposites could be sprayed in a wide variety of substrates including metals, paper and fabrics. Fig. 2(a) shows a photograph of this coating on a cotton fabric with water and oil droplets placed on it. On the other hand, the coatings obtained from the sprayed waterborne dispersion could be applied in a more limited variety of substrates due to the high temperature that the substrate had to maintain during the fabrication process. This was necessary, since without thermal treatment, due to the slow evaporation rate of water, dewetting effects would lead to coating inhomogeneities and cracks. These coatings were found to repel a wide variety of liquids as shown in Fig. 2(b). The table of Figure 2 shows the CA and ROA values for the different liquids.

4.0. Conclusions

We have developed two different formulations and spray-coating strategies in order to obtain superhydrophobic and superoleophobic coatings. The first formulation includes the use of a commercial biopolymer as a base material, mixed with hydrophobic nano-silica. A fluoroacrylic copolymer is used as a thin top-coat for surface chemical functionalization. The second formulation is obtained by spraying a waterborne fluoroacrlylic copolymer mixed with hydrophilic nano-silica. The fluoroacrylic copolymer used here is environmentally approved due to its C-6 chemistry. Such spray-based fabrication routes are important due to both their environmentally benign character and their ability to coat large surface areas.

5.0. References

1. Ragesh, P.; Ganesh, V.A.; Nair, S.V.; Nair, A.S.; J. Mater. Chem. A, 2014, vol. 2, p. 14773.
2. Song, W.; Veiga, D.D.; Custodio, C.A.; Mano, J.F.; Adv. Mater., 2009, vol. 21, p. 1830.
3. Wan, P.; Wu, J.; Tan, L.; Zhang, B.; Yang, K.; Mater. Sci. Eng. C, 2013, vol. 33, p. 2885.
4. Osawa, S.; Yabe, M.; Miyamura, M.; Mizuno, K.; Polym., 2006, vol. 47, p. 3711.
5. Milionis, A.; Ruffilli, R.; Bayer, I. S.; RSC Adv., 2014, vol. 4, p. 34395.
6. Higgins, C.P.; McLeod, P.B.; MacManus-Spencer, L.A.; Luthy, R.G.; Env. Sci. Technol., 2007, vol. 41, p. 4600.
7. Lau, C.; Thibodeaux, J.R.; Hanson, R.G.; Narotsky, M.G.; Rogers, J.M.; Lindstrom, A.B.; Strynar, M.J.; Toxicological Sci., 2006, vol. 90, p. 510.
8. Houde, M.; Czub, G.; Small, J.M.; Backus, S.; Wang, X.; Alaee, M.; Muir, D.C.G.; Env. Sci. Technol., 2008, vol. 42, p. 9397.
9. Bayer, I.S.; Davis, A.J.; Biswas, A.; RSC Advances, 2014, vol. 4, p. 264.
10. Milionis, A.; Dang, K.; Prato, M.; Loth, E.; Bayer, I.S.; J. Mater. Chem. A, 2015, vol. 3, p. 12880.

Electroforming Utilised for Replication of Large Scale Optical Surfaces

P. T. Tang*, M.Ø. Hansen, I. Mizushima
IPU, Produktionstorvet,
Building 425, 2800 Kongens Lyngby, Denmark
*Email:tt@ipu.dk

Abstract

Process chains for fabrication of tool inserts for injection moulding of optical polymer components are investigated. Two different optical components are studied; a backlight LED unit made by multiple embossing steps and a Fresnel lens made by diamond machining of an aluminium substrate. In both cases electroforming of nickel is utilised in order to replicate the delicate master and create a durable tool insert.

This paper describes and investigates the influence of each the required steps in the process chains, with respect to the replication fidelity and the resulting quality of the large scale polymer optics.

1.0. Introduction

Replication is perhaps the most important technique for bringing nanotechnology into real products at a reasonable price. Replication means that an extremely precise, expensive - and often also fragile - master can be accurately copied into a product, or into a tool that can be used to manufacture the product.

This paper focuses on the use of electroforming as a way to replicate a delicate optical master, and then utilise the resulting nickel tool for mass-production of polymer replicas. Two different optical applications are investigated; back light units for flat displays and Fresnel lens arrays for concentrated photovoltaics (CPV).

Since structures with an optical function are in the nanometre range almost by definition – dictated by the wavelength of visible light – large scale optics means any surface with an optical function consisting of numerous optical elements. This could be a relatively small part, such as a Blu-ray disc, or a much larger part consisting of several hundred replicated optical elements put together in an array (or any other type of order).

In the last years LED's have replaced the old CCFL tubes as light source for the back light unit in flat panel TV's. There are mainly two types of LED backlight units which are common. First, there is the so called direct type of LED back light unit, where an array of multiple LEDs is placed behind a diffuser plate. This approach has some benefits concerning local dimming functionalities and brightness. Secondly, there is the indirect type, which makes use of a light guide plate. LED light is coupled into one side of the light guide plate. The light propagating through the light guide plate is out-coupled homogenously by means of patterns (optical micro structures) on the light guide surface. This approach allows for very flat designs and economic production of the back light units. However, the performance of the state-of-the-art light guide plates is insufficient. In conventional indirect type back light units (see Fig. 1), screen printed or laser patterned light guide plates are used. Those methods just enable diffused out-coupling of the light. In order to achieve the highest degree of efficiency and light direction, complex micro-optics needs to be manufactured into the entire surface (up 65" or more).

Similarly large scale optics are also required for applications such as retro reflectors and solar collectors based on Fresnel lenses[1]. Today these applications require both high resolution and aspect ratios of structures, and prices are going up dramatically with the size of the area that the optics must cover[1]. Machining of optical structures is expensive, and hence replication from a master and parquetting of the replicas is the preferred way to create large scale optical tools[1].

As mastering process, diamond turning is the most appropriate technique to make optical nano-structures, compared to other techniques such as photolithography and laser processes - which suffer from limited resolution, structure sharpness and overall uniformity. The problem with diamond turning is that only a few substrate materials are available, since the substrate must be hard enough so that the generation of burrs at the facets is minimized, and at the same time without causing wear to the diamond tool itself. Typically used materials for diamond turning are electroless nickel, fine grained copper or aluminium and various types of bronze and brass. However, only electroless nickel has the mechanical strength and hardness to function as a tool during injection moulding – and the cost of electroless nickel can become a critical factor for production of large scale optics.

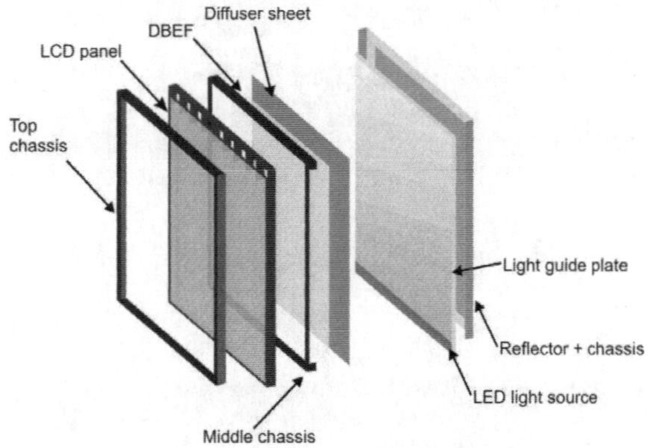

Fig. 1: LCD television set-up using the indirect type of back light unit. A more homogeneous and more directed light out coupling would allow for substitution of the diffuser sheet and of the brightness enhancement films (DBEF) [3].

Hence the search for alternative process chains for replication of large scale optical structures is the main focus of this paper.

2.0. Process chains

Several process chains and variants will lead to replication of a fragile nano-structured master. In this paper two specific process chains will be investigated, both based on nickel electroforming (Figs. 2 and 3).

The first process chain is shown in Fig. 2. Nickel is deposited by electroplating on a master characterized by a structured surface fabricated by diamond machining. The nickel deposit is separated from the master, and the master reused for several more replication steps. Finally the nickel replicas are placed together (so-called parquetting) to form a larger area.

In the second process chain an embossing tool is carefully produced using diamond turning and mounted in the embossing machine. A blank master plate made of a suitable substrate material (to be able to make on-line optical measurements usually a transparent polymer material is chosen) is then structured by the tool using adaptive step-and-repeat embossing[3]. This process will take a considerable amount of time, however - since the master plate is replicated into nickel and then used for mass-production by injection moulding - this high cost can be accepted.

2.1. Fabrication of nano-structured master

The process chain chosen for production of the back-light units is remarkably clear and straight forward. A transparent polymer plate will be structured by thousands of accurate embossing steps, during which various heated tools are pressed into the substrate to create small optical structures. After embossing has been completed the replication process starts, with the goal of turning the soft polymer master into a durable metallic tool that can be utilized for injection moulding of thousands of products – each of them an accurate copy of the original polymer master.

It is extremely important however, that the process chain used to replicate the master is not corrupting the fine structures master and that the low surface roughness of the master is also replicated.

Using a transparent polymer material for embossing the master plate, it is possible to use an in-line metrology system to constantly measure the quality and optical output of the master.

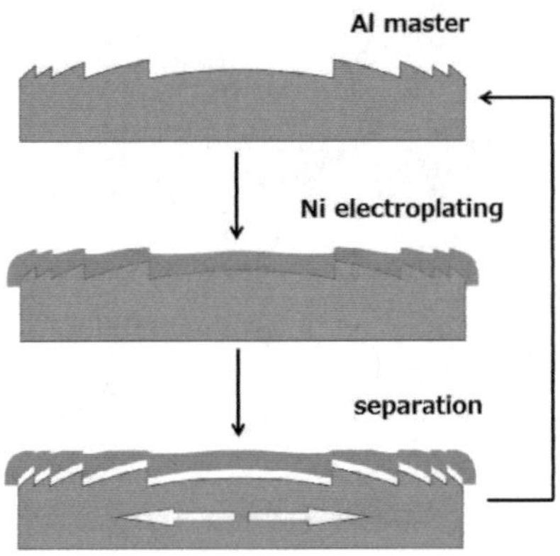

Fig. 2: Schematic drawing of the process chain for replication of Fresnel lenses diamond machined into an aluminium substrate.

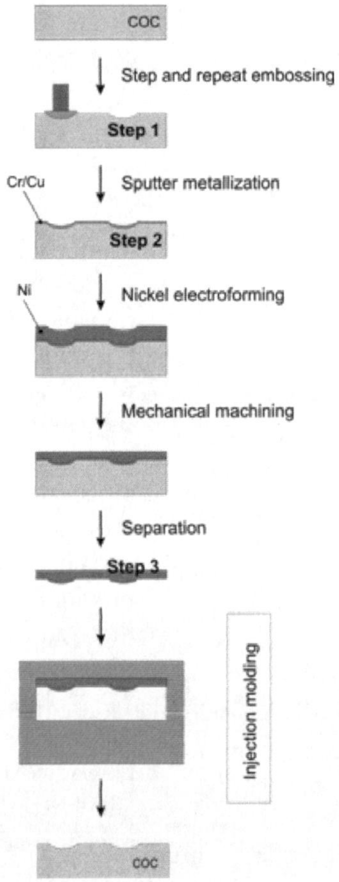

Fig. 3: Simplified drawing of the process chain used to replicate a polymer master structured repeatedly using embossing of a simple lens shape. Steps 1 through 3 correspond to the metrology scans performed on a small portion of the lens array.

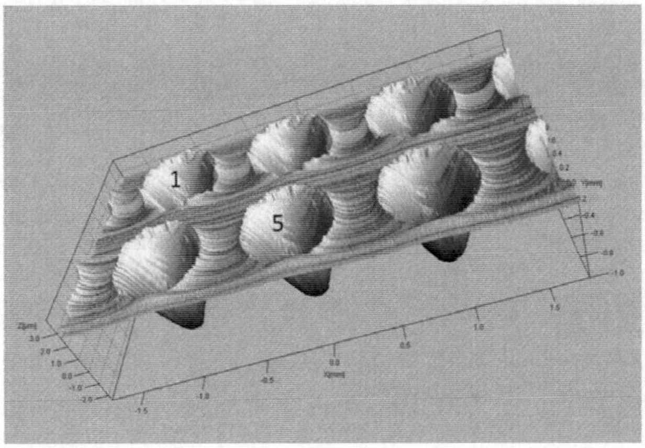

Fig. 4: Metrology scans of the 6 lenses in the upper left hand corner of the lens array, with indication of lens number 1 and 5. Please note that the scale for X and Y are in mm, while the scale for the Z axis is in μm.

Partly as a result of this, it is an advantage if the substrate is made of the same material as the final product - and among the possible transparent thermoplastic materials, such as PMMA, COC, PC, PS or PET, cyclic olefin co-polymer (COC) was chosen as substrate for this investigation. The substrate was then embossed using an adaptive step and repeat process [3]. The overall size of the master was 130 x 122 mm with a thickness of 6.35 mm. The COC materials was TOPAS 6013 kindly supplied by Topas-US.com. The stamp used for the step and repeat embossing process, was a simple lens structure as roughly illustrated in Fig. 3. After filling the substrate with identical lenses (excluding the edges), the surface was subjected to metrology. The surface scan was recorded as **Step 1** (see Figs. 3 and 4). Only the 6 lenses in the upper left hand corner were scanned.

Similarly for the concentrated photovoltaics, a single master was produced by diamond machining of a fine-grained aluminium substrate. Each master plate is replicated into nickel several times.

2.2. Pre-treatment of the master

For the Fresnel lenses machined in aluminium, it was important to find a good pre-treatment of the substrate. Several different methods to increase (or reduce) the thickness of the natural oxide layer on the surface were tested. However, it turned out that as long as the surface is clean the adhesion to the natural oxide of aluminium is sufficiently weak for the separation to work (and not so weak the delamination occurs during electroforming). To clean the surface a simple procedure of rinsing the substrate with water and soap, rinsing the water, cleaning with ethanol and rinsing with water was used.

For the COC master plate several metal layers are deposited on the transparent polymer after embossing. The main challenge is to deposit a thin coherent layer, sufficient for providing the electrical conductivity required for the following galvanic processes. On the other hand the metallization layer must not alter the optical features of the polymer master by thermal decomposition or uneven material distribution. The best results have been obtained using magnetron sputtering of chromium and copper.

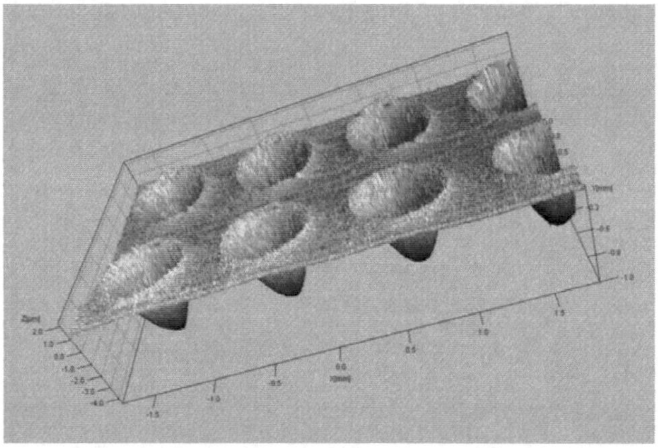

Fig. 5: Metrology scans of the same area of the plate after the metallization with chromium and copper. Comparing with fig. 4 it is obvious that the surface roughness has increased.

Chromium provides adhesion to the polymer substrate and copper provides conductivity for the electroforming (galvanic) step. For this investigation 60 nm of chromium and 200 nm of copper was sputtered onto the surface of the COC master. Prior to magnetron sputtering the COC master was plasma cleaned in a mixture of argon and residual gas in the chamber at $70 \cdot 10^{-3}$ mbar. The magnetron sputtering was conducted in argon plasma at $10 \cdot 10^{-3}$ mbar. After magnetron sputtering another surface scan was performed, called **Step 2** (see Fig. 5).

2.3. Galvanic deposition

The first galvanic layer will, when the tool insert is completed, become the surface of the injection moulding tool (ignoring any prior metallization layers since they are extremely thin and might be removed during the separation step). For this reason the galvanic layer must be wear resistant and hard enough to with-stand the high pressures of the injection moulding process.

At first the galvanic process must progress relatively slowly since the electrical conductivity of the master is low due to the thin metallization layer (this is not a problem in case of an aluminium substrate). Later on the current density can be increased, in order to reach the desired thickness in a reasonable time.

To avoid thermal stress, induced by the large difference in thermal expansion coefficient for the various materials (polymer, aluminium, metallization layers and galvanic layer), the process temperature during electroforming should be as low as possible – preferably close to room temperature. So far the best results have been obtained using sulphamate nickel electrolytes. Three different electrolytes have been established operating at either 32 [4], 40 [4] or 50 °C [5] and with plating rates from very slow (15 μm/hour) to very fast (200 μm/hour).

Due to the relatively poor adhesion between the deposited nickel and either the metallization layers (Cr/Cu) and the substrate (COC) or the aluminium substrate, it is of the out-most importance to keep the internal stress in the electroplated nickel as close to zero as possible. Any deviation from a close-to-zero stress behaviour will result in de-lamination of the outer areas of the nickel deposit (particularly in corners and edges), and defoliation in the middle area of the part. This effect is created by the fact that the internal stress in sulphamate nickel deposits (however small they may be), will depend on the local current density – so that a low level of tensile stress will exist at the perimeter of the part (corners and edges) while a low level of compressive stress is found in the middle of the part [6].

For this investigation approximately 300 to 1000 μm of nickel was deposited on top of the metallization layers (or the diamond turned aluminium substrate), using a relatively low current density of 1.38 A/dm^2 – corresponding to a nickel deposition rate of 15 μm/hour.

In case of the Fresnel lenses the most challenging issue is to control the adhesion between the nickel deposit and the master. The adhesion must be good enough that the nickel is not delaminated during electroplating and poor enough so that the nickel can be separated from the master without losing accuracy of the surface microstructure. The adhesion is influenced by the master material, residual stress of the nickel deposit, the optical structure machined and thermal mismatch between the deposit and the master. More information on measuring and controlling the internal stress in the electroplated nickel can be found in [7].

In order to measure the required force for separation of the nickel deposit from the aluminium master, special aluminium substrates and holders were made for the nickel electroplating and for the separation test. The master is a round aluminium plate, 5 mm in thickness and 70 mm in diameter. Before nickel electroplating, the aluminium plate was rinsed with soap and ethanol and placed in the holder. A conductive cable was connected to a screw attached to the back of the plate through the holder (this way deposition on the back of the substrate is prevented by the plastic holder). Nickel was deposited on the plate by electroplating for 24 hours. The thickness of the deposit was approximately 300 μm. To further reduce thermal stress due to mismatch of the thermal expansion between nickel and aluminium, the plate with the holder was maintained at the same temperature as the electrolyte (32 °C) after electroplating.

2.4. Separation of replica and master

For the lens structures embossed in COC, the most gentle and least damaging way to separate master and replica is to dissolve the COC master in a solvent that does not – in any way – attack the surface of the replica. However, the easiest way is to separate the nickel and master mechanically. In this investigation the mechanical method was used, and immediately after the separation both the nickel replica (the scan called **Step 3**) and the COC master (this scan is called **Step 1a**) was measured using the profilometer.

For the Fresnel lenses in aluminium, 8 screws were placed in pre-machined holes in the back of the aluminium substrate after electroforming and then placed in a holder for separation testing. The holder consists of 3 plates, and the nickel plated aluminium was sandwiched between two of them. During electroforming the diameter of the nickel deposit grows a little and thus becomes larger than the Al plate due to deposition at the edge. The nickel deposit can then be separated by pulling the 8 screws while holding the nickel deposit in the circular edge. The separation test was performed with a universal tensile testing machine (hydraulic AMSLER).

2.5. Mechanical machining

For large scale electroforming it is impossible to deposit a completely uniform layer of nickel across the diameter of the substrate. Furthermore, since the nickel electroforming copies the features of the substrate, the features will still be visible even after deposition of several millimetres of nickel[8]. Thus some kind of mechanical machining (milling, polishing, grinding, etc.) is almost always required in order to obtain the flat surface of the back of the tool insert that the injection moulding needs in order to make perfectly even parts. Laser cutting or wire EDM is used to cut the nickel replica to the requested size for the tool insert. This cutting process is particularly important for the Fresnel lenses, since several nickel replicas are placed together in an array (called parqueting) to from the complete tool for injection moulding. Consequently, any error in the size of the nickel pieces or in quality of cut edges will result in a visible line in the finish polymer product.

Since the wear resistance of the Cr/Cu metallization layers used for the back-light unit is not expected to be very good, and thus would slowly be removed during injection moulding, the metallization layers were dissolved using a proprietary ammonia based solution which is known not to attack nickel[9]. After dissolving the metallization layers the nickel replica was measured again using the profilometer and this scan is called **Step 3a** (see Table 1).

It was not necessary to clean or alter the nickel tools made by electroforming of Fresnel lenses in aluminium, however wire electro-discharge machining (wire-EDM) was used to accurately cut the nickel shims into perfectly rectangular tool inserts (to be used to parquetting of larger areas).

2.6. Injection moulding

The final goal of the process chain is the injection moulding. Here the tool insert - which has been fabricated in a durable nickel material - is used to reproduce the optical features into (hopefully) thousands of replicas.

It is beyond the scope of this investigation to look into the various parameters of the injection moulding process, and their effect on the replication accuracy. However, obtaining good replication of the tool insert is also not a trivial issue – and will almost always involve a trade-off between cycle time and replication quality. This is due to the fact that a pre-heated tool surface will be replicated more accurately by the progressing melt front of the polymer during injection – while at the same time a heated tool insert will require longer time for cooling, before the finished part can be ejected. Some recent experiments with induction heating of tools for injections moulding[12], have demonstrated that if carefully optimised heating can be applied almost without increasing the cycle time.

3.0. Methods and results

3.1. Metrology of the back-light unit

To be able to compare the surface quality, and later on the replication quality, a number of scans were performed. The instrument used was a mechanical 50 mm inductive profilometer (Talysurf series 2, by Taylor Hobson) with a 90° conical stylus and a spherical tip (2 μm radius). The probing force was nominal 1 mN (static measuring force 0.7 mN). The vertical measuring range was 0.200 mm with a 3 nm resolution. The data resolution in horizontal traverse was 0.25 μm.

Data for the 3D scans were collected by data logging a number of parallel profiles at a pre-determined trace length and spacing using a Y-axis stage (range 50 mm and 1 μm resolution). During measurements a resolution equal to 2 μm along the X axis and 10 μm along the Y axis, with a measurement range of 3.5 x 2.0 mm, was obtained. The measurement speed was 1.0 mm/s.

All measurements were performed on the 6 lenses in the upper left hand corner of the lens array.

The tool used for the embossing of the lenses was not perfect, and scratches – originating from the tool – are found in specific areas of all the lenses in the lens array. For the measurements this is actually helpful, since it makes it easier to identify the same position of the scans performed at various steps of the process chain.

Fig. 6: (a) Nickel deposit which was broken at the edge during the separation from the substrate and **(b)** nickel deposit and aluminium substrate after successful separation test using a load of 6850 N.

Table 1: Calculated roughness (Sa) focusing on an almost equally sized rectangular area inside the lenses (in μm). The steps correspond to the process steps illustrated in Fig. 3. Step 1a is the COC substrate after mechanical separation of the nickel, step 3a is the nickel replica after the metallization layers have been selectively removed.

Step	1	2	3	1a	3a
Lens 1	0,036	0,061	0,035	0,037	0,050
Lens 5	0,033	0,060	0,042	0,038	0,049

To visualize the profilometer scans, and to measure the height and calculate the surface roughness, the software SPIP Scanning Probe Image Processor (version 5.1.4 from Image Metrology A/S [10]) was used.

To calculate the roughness of the inside of the lenses, a rectangular area was selected (the largest possible square that will fit inside the circular perimeter of the top/bottom of each lens). Then a line wise correction of 3 degrees was performed by the software, to correct for the parabolic shape of the lens. From the result of this correction, the central part was selected (to avoid the scratched area found in every lens due to a surface error of the embossing tool) and used for calculation of the Sa value (see Table 1).

3.2. Metrology of the Fresnel lens

Table 2 shows the results of several different separation tests. For the first test the nickel deposit was broken at the finest structure on a load of 1230 N upon separation test (Fig. 6(a)). The thickness of the deposit of about 300 μm was not enough to withstand the force for separation, partly because a remaining burr from the machining process was detected during a line scan of the master. For the second test a deposit three times as thick as the first sample was separated with a load of 6850 N. The replication seems to have succeeded and the micro structured surface was visually fine. In order to see if the micro structure of the master surface could withstand the separation process, the geometry of the structure was measured using a stylus profilometer (Taylor Hobson: series 2, 50, provided with a 2 μm radius tip). The used resolution was 0.25 μm, and the measurements were carried out scanning a single profile from the edge to the centre in 4 different places. The analysis was performed using the software Scanning Probe Image Processor (SPIP) Ver. 5.1.4.

Table 2: Loads required for separation of the nickel deposit from the aluminium substrate. The internal stress was measured (see [7]) in the electrolyte before and after electroforming which takes from 24 to 72 hours.

Internal Stress in nickel deposit (MPa)		Thickness (μm)	Heating during separation	Load (N)
before	after			
-41	-70	about 300	-	>1230
0	-70	about 900	-	6850
15	-60	about 900	+	1920
0	0	about 900	+	1110

The profiles show that the micro structure of the master was actually damaged by the separation force (comparing Fig. 7(a) and 7(b)) for experiment number two in Table 2.

To further reduce the required force for separating master and nickel replica, the materials were heated to 50 °C during the separation. This way the higher thermal expansion of aluminium (compared to nickel) helps to perform the separation, since the substrate expands more than the deposited nickel (indicated by the arrows in Fig. 2). After introduction of heating during separation, the required force drops from 6850 N to 1920 N (see Table 2). An additional fine tuning of the internal stresses in the nickel deposit (last experiment in the Table 2) further reduces to required force to 1110 N. Both of the two experiments performed with heating during separation, resulted in good undamaged aluminium substrates – which could be used again to create another nickel replica.

4.0. Discussion

4.1. Embossing into COC

The overall process chain, as illustrated in Fig. 3, seems to function in the sense that it is possible to obtain a tool insert with an acceptable quality. This insert can then be used for injection moulding of accurate copies of the original master, although unfortunately it was not possible to include measurements of injection moulded parts in this investigation.

The sputter metallization layers provides the necessary "controlled poor adhesion" over the entire surface of the master and makes the surface electrically conducting for the following galvanic deposition of nickel.

The biggest reduction in quality – or the biggest deviation from the surface of the master – occurs during the sputter metallization of the embossed COC substrate. This is relatively clear when comparing Fig. 4 and Fig. 5, and looking at the Sa-values in table 1 there is a significant increase in roughness moving from **Step 1** (the COC master) to **Step 2** (the metalized master).

Nickel was deposited in a virtually stress-free mode, at a low temperature (32 °C), making sure that stresses originating from thermal expansion mismatch does not become critical.

Using relatively flat and smooth optical structures, such as an array of lenses, it is possible to separate the nickel insert from the COC master in a non-damaging way.

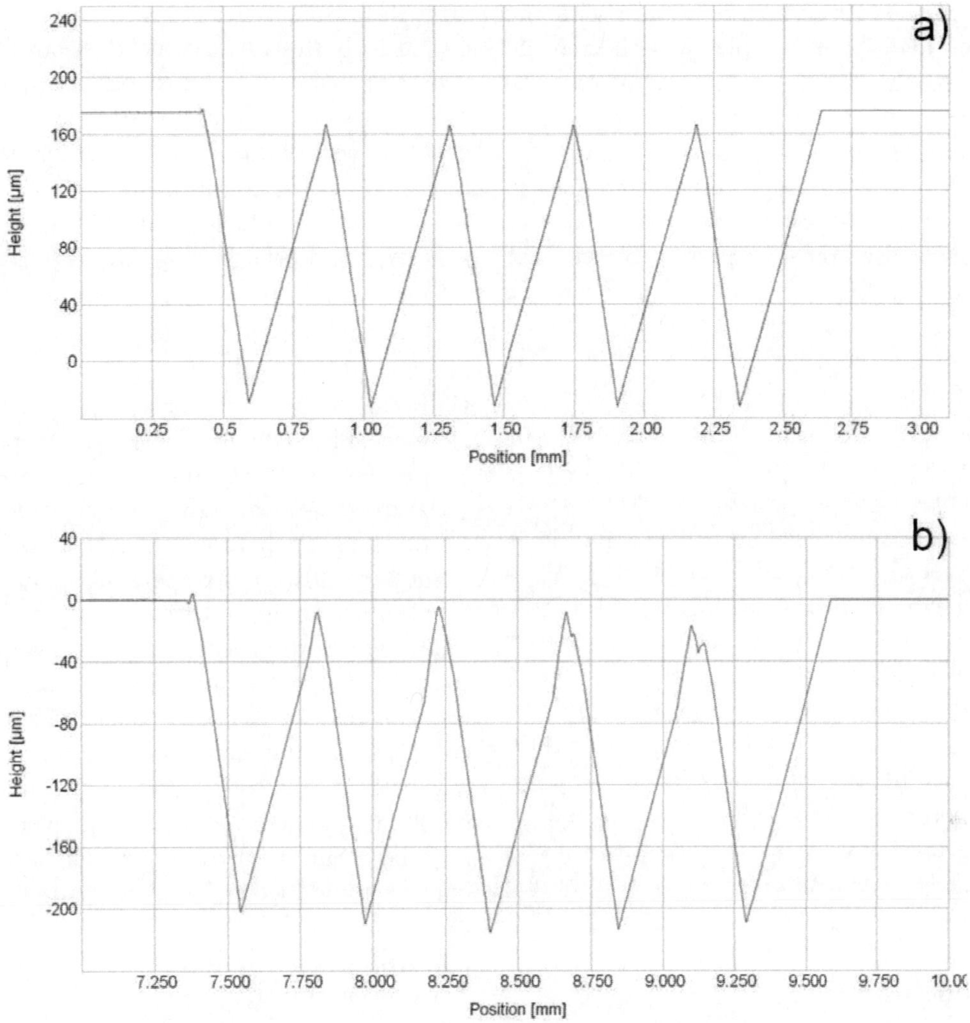

Fig. 7: (a) Geometric profiles of the aluminium master before and **(b)** after separation test (corresponding to exp. 2 in table 1) using a load of 6850 N.

This opens the possibility of reusing the master. However, from other investigations[11] it appears that something remains in the lenses after the mechanical separation of the COC master and the metallic replica, since the depth of the all the lenses for **Step 1** are larger than for **Step 1a**. The roughness however, seems not to be changed significantly.

Finally, it can be concluded that the best tool insert is obtained if the metallization layers (chromium and copper) are not removed from the underlying nickel. However, since the wear resistance of copper is not very high – and the metallization layers will become the outermost surface in the injection moulding tool – it would be better to find a combination of metallization layers with a higher wear resistance. This way it would also not be necessary to remove the layers (leading to **Step 3a**), and roughness values would be as **Step 3** in tables 1 – which are clearly closest to the COC master (**Step 1**).

4.2. Fresnel lenses

In order to enable separation of aluminium master and nickel replica with the smallest load, the sample was heated to approximately 70 °C during the mechanical separation. The thermal expansion coefficients of aluminium and nickel are 23.7 and 13.0 μm/m·°C respectively. This difference in thermal properties can be utilized to reduce the force for the separation, since the expansion of the aluminium plate push the nickel deposit out (following the curved surface of the Fresnel structure).

In this case samples were kept at approximately 70 °C by a heating gun. The nickel was separated with a load of 1920 N, less than 30% of the force required for separation without heating (for a comparable sample, see also exp. 2 and 3 in Table 2).

High internal stress in the nickel gave rise to corrosion of the aluminium master upon electroforming due to the (intended) poor adhesion between the substrate and the nickel deposit, since the stress resulted in partial delamination of the deposited nickel. Stress control of the nickel was achieved by keeping the concentration of all chemicals in the nickel electrolyte constant – also during electroforming.

5.0. Conclusion

A replication process for fabrication of tools for large scale polymer optics using a master in either aluminium or transparent polymer (COC) has been demonstrated. In both cases it is possible to replicate the master and transfer the structures in the surface of the master to the replica. The replica is made by electroforming and the resulting nickel part can be used as tool insert in an injection moulding process.

Two process chains have been investigated; one in which a small diamond machined tool is embossed into the surface of a large polymer master and one in which a diamond machined aluminium master is replicated several times and the replicas are placed next to each other to form a larger array of Fresnel lenses.

In both cases the nickel electroforming process provides an almost perfect replication of the master. However, looking at the entire process chain the largest contribution to replication errors comes from;

- For the embossed COC master an increased surface roughness resulted from the magnetron sputtering used to provide electrical conductivity before the electroforming process
- For the aluminium master, large mechanical forces during the separation process could in some cases damage the sharp edges of the Fresnel structure – particularly when the separation takes place at a lower temperature than the electroforming process

For both process chains it was possible to separate master and nickel replica with a minimal damage to the master, which is the most important requirement for reuse of the master (and thus for reducing cost). However, it will only be possible to reuse a given master a limited amount of times due to wear, corrosion and mechanical deformation.

6.0. Acknowledgement

The authors are grateful for the expert assistance regarding metrology measurement and the SPIP software provided by Stefania Gasparin and René Sobiecki from the department of mechanical engineering at the technical university of Denmark (DTU). The authors are also grateful to Christoph Baum from Frauenhofer-Institut für Produktionstechnologie (IPT) in Germany and to Henrik Madsen from MillPond Optics (Denmark). Finally the authors wish to thank Christian Ravn and Klaus S. Hansen from IPU for their assistance and the European Union for supporting the FP7 research project FlexPAET (Flexible Patterning of Complex

Micro Structures using Adaptive Embossing Technology) and the Research for the benefit of SME'S project PolyGlass (Development of a new method to produce high efficiency solar concentrators based on polymer casted directly on glass).

7.0. References

1. Scott, S.; Diffractive Optics and Micro-Optics, Optical Society of America Technical Digest, paper DTuD4, 2004.
2. Riemer, O.; Key Eng. Mater., 2007, vol. 364-366, p. 13.
3. Baum, C.; et al.; Int. Semicond. conf. CAS, Oct. 10-13, Sinaia, Romania, 2010.
4. Proprietary sulphamate electrolytes developed by IPU.
5. Correspondence between the authors and O. Humbach from the company Temicon GmbH, Germany
6. Tang, P.T.; Benzon, M.E.; Fontenay, F.S.; Rasmussen, J.P.; SUR/FIN '96, 1996.
7. Mizushima, I.; Tang, P.T.; Hansen, K.S.; Gasparin, S.; Madsen, H.; Proceedings 9th Int. Conf. Multi-Material Micro Manufacture(4M), 2012, p. 85.
8. Tang, P.T.; 4M Conf., 2008, p. 18.
9. Tang, P.T.; "A Method Of Manufacturing A Mould Part", EP1807553, US 7,767,263 B2 (issued August 2, 2010)
10. www.imagemet.com
11. Tang, P.T.; Sobiecki, R.; Hansen, M.; Ravn, C.; Baum, C.; Proceedings ICOMM, 2011, p. 431.
12. Menotti, S.; Hansen, H.N.; Bissacco, G.; Calaon, M.; Tang, P.T.; Ravn, C.; Int. J. Adv. Manuf. Technol., 2014, vol. 74, p. 907.

On Influence of Electrolyte Flow Rate and Rotary Speed on Surface Modification of Bevel Gear Finished by PECH

Sunil Pathak, N.K. Jain*, I.A. Palani
Discipline of Mechanical Engineering,
Indian Institute of Technology Indore (MP) India.
*Email: nkjain@iiti.ac.in; nkjain73@yahoo.co.in

Abstract

Surface quality of bevel gears namely surface finish and micro-geometry has great influence on their service life, operating performance and noise generation. Pulse electrochemical honing (PECH) process is developed with an aim to improve both aspects of surface quality of the bevel gears in a single operation. The main objective of the present work is to investigate the influence of electrolyte flow rate and rotary speed of the workpiece gear on the surface quality of bevel gears finished by PECH. Pre-identified values of electrolyte composition (as 75% $NaNO_3$ + 25% NaCl), electrolyte concentration (as 7.5 wt. %), pulse-on time (as 2 ms), pulse-off time (as 4.5 ms) and finishing time (as 6 minutes) were used for present work. Surface roughness of the bevel gear was studied in terms of percentage improvements in average surface roughness, maximum surface roughness and depth of surface roughness. Microgeometry of the best finished bevel gear was studied in terms of errors in pitch and runout. Microstructure of the best finished gear was also studied. The results have shown significant improvements in the measured responses.

1.0. Introduction

Pulsed electrochemical honing process (PECH) combines the benefits of pulsed electrochemical finishing (PECF) and mechanical honing process and simultaneously overcomes their limitations in a single operation to improve surface quality of the bevel gear tooth profile. Material removal in PECH is based on a controlled anodic dissolution of workpiece (anode) with the tool as the cathode in an electrolytic cell. Some researchers in past[1–5] have tried to overcome these limitations by using electrochemical honing (ECH) for finishing different types of gears.

Shaikh and Jain[6] envisaged a novel concept of complementary cathode gears to meet challenges of straight bevel gear finishing by ECH. They also developed mathematical models for material removal rate and surface finish of ECH-finished bevel gears[7]. Pathak et al.[8] reported on using PECH for improving micro-geometry of straight bevel gears. Pathak et al.[9] compared the process performance of ECH and PECH for bevel gear finishing and concluded that PECH gives better surface quality and micro-geometry of the bevel gears than ECH but PECH is less productive than ECH. The objectives of the present work is finish the bevel gears by PECH and optimize the values of electrolyte flow rate and rotary speed of the workpiece gear (i) to minimize average surface roughness 'R_a', maximum surface roughness 'R_t', depth of surface roughness 'R_z'; and (ii) to minimize the errors in single pitch error 'f_p', adjacent pitch error 'f_u', cumulative pitch error 'F_p' and total runout 'F_r'.

2.0. Methods

Eight experiments were conducted to study the influence of electrolyte flow rate and rotary speed of workpiece gear on surface finish and microgeometry of straight bevel gears by varying electrolyte flow rate 'F' at three levels and rotary speed of workpiece gear 'R' at four levels using D-Optimal technique of response surface methodology. Pre-identified values of electrolyte composition (as 75% $NaNO_3$ + 25% NaCl), electrolyte concentration (as 7.5 wt. %), pulse-on time (as 2 ms), pulse-off time (as 4.5 ms) and finishing time (as 6 minutes) were used during the experimentation. The experiments were conducted on the experimental setup developed for bevel gear finishing by PECH by Pathak et al.[8]. This setup has four major units namely (i) arrangement of workpiece, cathode and honing gears in a machining chamber as shown in Fig. 1, (ii) unit for supplying, cleaning and recirculating the electrolyte at a desired temperature, pressure and flow rate; (iii) pulse-DC power supply system; and; (iv) provision for providing rotary motion to the workpiece gear and an automated feeding motion for engagement and disengagement of workpiece gear with the mating cathode and honing gears. The workpiece gear and honing gear are made of 20MnCr5 alloy steel due to their overwhelming demand in the field of automobile and aerospace engineering.

Three parameters of surface roughness (R_a, R_t and R_z) were measured using the contracer-cum-surface roughness analyzer (model SEF 3500) from *Kosaka Japan* using cut-off length of 0.8 mm and evaluation length of 4 mm. Three measurements were taken at different locations on middle of the left and right hand flanks of two consecutive gear teeth. These 12 values were used to calculate arithmetic average of a surface roughness parameter of an unfinished gear and same gear finished by PECH. Percentage improvements in average surface roughness (R_a) value 'PIR_a' was calculated using eq. 1. Similarly, average values of PIR_t and PIR_z were computed using their corresponding average values.

Fig. 1: Photograph of machining chamber for bevel gear finishing by PECH

A higher value of percentage improvement in a surface roughness parameter implies lower value of that parameter after finishing by PECH.

$$Avg.\ PIR_a = \frac{Avg.\ R_a\ value\ of\ unfinished\ gear - Avg.R_a\ value\ of\ finished\ gear}{Avg.R_a\ value\ of\ unfinished\ gear} \quad (1)$$

Microgeometry of the bevel gears was evaluated in terms of four parameters namely single pitch error (f_p), adjacent pitch error (f_u), cumulative pitch error (F_p), and total runout (F_r) were measured before and after finishing by PECH using a computer numerically controlled (CNC) gear metrology machine *SmartGear 500* from *Wenzel GearTec, Germany.*

3.0. Result And Discussion

3.1. Analysis of Surface Roughness

Table 1 presents values of the input parameters and average values of surface roughness parameters along with average percentage improvements in them for all the experiments. It can be observed from these results that the maximum values of PIR_a, PIR_t and PIR_z (62.45 %; 60.58%; and 53.98%) were obtained in experiment no. 3 using electrolyte flow rate as 20 lpm and rotary speed of workpiece gear as 40 rpm.

Fig. 2(a) and Fig. 2(b) depict the variation of average values of PIR_a, PIR_t and PIR_z with electrolyte flow rate and rotary speed of the workpiece gear. In these graphs, average value of a response at particular level of input parameter has been shown. It can be observed from these graphs that lower value of electrolyte flow rate (i.e. 20 lpm) and an optimum value of rotary speed (i.e. 40 rpm) yield the maximum values of *avg.* PIR_a, *avg.* PIR_t, and avg. PIR_z. These trends can be explained as follows:

a) Lower electrolyte flow rate provides the electrolyte sufficient time to build up the required ion concentration and for allowable electrolyte temperature rise which are required for the electrolytic dissolution of the anodic workpiece material. This results in better finished surface for given finishing duration i.e. higher values of percentage improvements in *avg.* PIR_a, *avg.* PIR_t, and *avg.* PIR_z. Whereas, higher electrolyte flow rate leads to more amount of fresh electrolyte which maintains electrolyte temperature almost constant.

Table 1: Input and surface roughness parameters of bevel gears for different experimental runs

Input variables			Average values of surface roughness parameters and average percentage improvements in them								
Run no.	Electrolyte flow rate (lpm)	Rotary Speed (rpm)	*Avg.* R_a before PECH	*Avg.* R_a after PECH	Avg. PIR_a	*Avg.* R_t before PECH	*Avg.* R_t after PECH	Avg. PIR_t	*Avg.* R_z before PECH	*Avg.* R_z after PECH	Avg. PIR_z
1	40	60	2.14	1.95	8.88	18.4	16.1	12.50	11.6	8.9	23.28
2	20	60	1.5	1.1	26.67	18.1	15.2	16.02	8.76	6.95	20.66
3	***20***	***40***	***2.69***	***1.01***	***62.45***	***24.1***	***9.5***	***60.58***	***20.1***	***9.25***	***53.98***
4	30	60	2.85	1.8	36.84	18.1	12.2	32.60	15.7	12.2	22.29
5	20	20	2.18	1.05	51.83	19.1	11.2	41.36	14.5	8.75	39.66
6	40	20	2.08	1.95	6.25	24.2	18.25	24.59	15.5	15.1	2.58
7	30	30	1.71	1.2	29.82	13.1	8.95	31.68	9.5	6.2	34.74
8	40	40	1.56	1.4	10.26	13.7	12.5	8.76	8.47	7.2	14.99

This decreases ion concentration which reduces finishing action and results in higher values of surface roughness parameters of the finished surface i.e. lower values of *avg.* PIR_a, *avg.* PIR_t, and *avg.* PIR_z.

b) Very high rotary speed provides lesser amount of time per cycle available for finishing of workpiece gear surface by electrolytic action and increases amount mechanical honing. This results in less amount of material being removed from the surface peaks which yields lower values of percentage improvements in surface roughness parameters. Whereas, very low rotary speed provides excessive time for electrolytic dissolution of the workpiece gear surface which also yields poor surface finish due to selective dissolution of workpiece gear material, sporadic breakdown of anodic film, flow separation and formation of eddies, and evolution of hydrogen gas. Therefore, an optimum value of rotary speed exists corresponding to optimum combination of electrolytic dissolution and mechanical honing.

3.2. Analysis of Microgeometry

Errors in pitch and runout the prime reasons for poor tooth contact and improper meshing and are responsible for generation of noise and motion transfer errors. Fig. 3(a) and Fig. 3(b) present the reports generated by the CNC gear metrology machine for the best finished gear i.e. experiment no. 3 for single pitch error and adjacent pitch error of the straight bevel gear before and after its finishing by PECH. Figs. 4a and 4b present the same for the cumulative pitch error and total runout. It can be inferred from the Figs. 3a-3b that percentage improvement in single pitch error (PIf_p) and percentage improvement in adjacent pitch error (PIf_u) for left hand flank are 47.3% and 41.1% respectively. This improves the quality of the gears in *Deutsche Normen(DIN)* standard from DIN standard 9 to 7 for single pitch error 'f_p' and from DIN standard 8 to 6 for adjacent pitch error 'f_u' of the bevel gear. From Fig. 5, the percentage improvements obtained in cumulative pitch error (PIF_p) and in total runout (PIF_r) for left hand flank are 44.7 % and 40.7% respectively. This improves the gear quality from DIN standard 8 to 6 for cumulative pitch error 'F_p' and from DIN standard 7 to 6 for total runout 'F_r'.

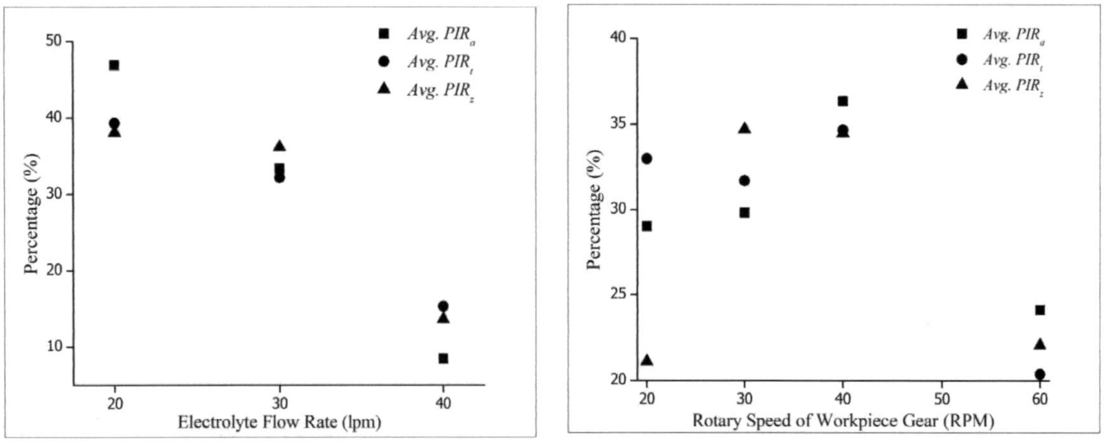

Fig. 2: Variation of percentage improvements in average surface roughness *(PIR_a)*, in maximum surface roughness *(PIR_t)*, and in depth of roughness *(PIR_z)* with **(a)** electrolyte flow rate **(b)** rotary speed of workpiece gear

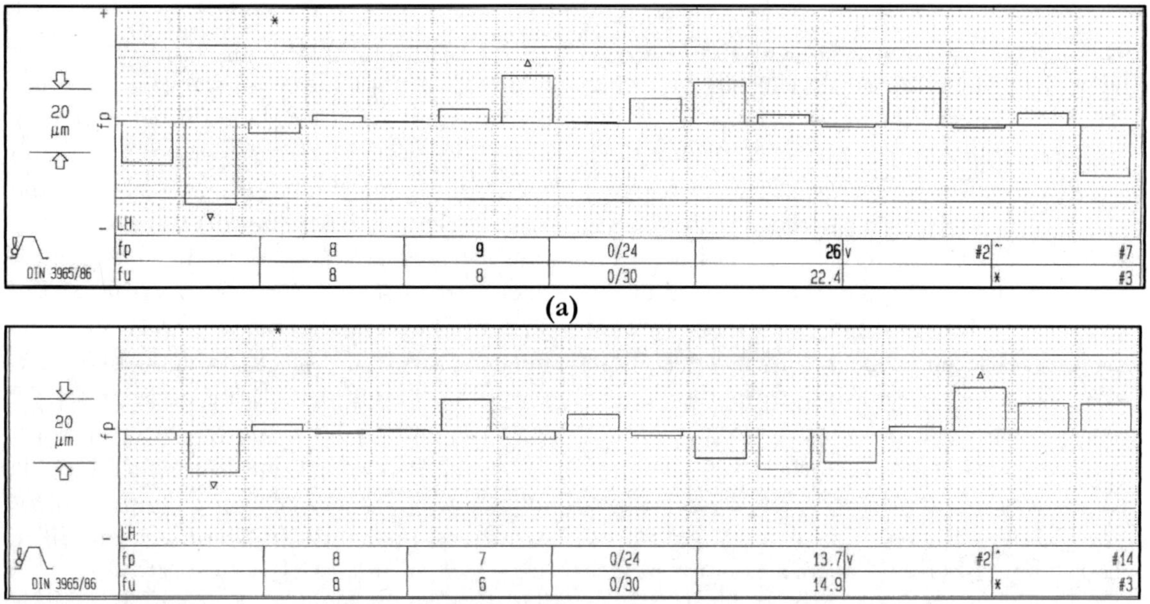

Fig. 3: Single pitch error 'f_p' and adjacent pitch error 'f_u' of bevel gear tooth for the best finished gear i.e. experiment no. 3 **(a)** before PECH and **(b)** after PECH.

3.3 Analysis of Microstructure and Microhardness

Figs. 5(a) and 5(b) depicts the SEM images of the unfinished and finished bevel gear tooth by PECH at identified optimum input parameters at 400x magnification. It was observed form these figures that the cutting marks and rough surfaces which are generated on the gear tooth flank during their manufacturing are completely removed from the PECH finished gear. Also, the micro-pits in the gear material which was not visible in unfinished gear tooth are clearly visible after finishing by PECH.

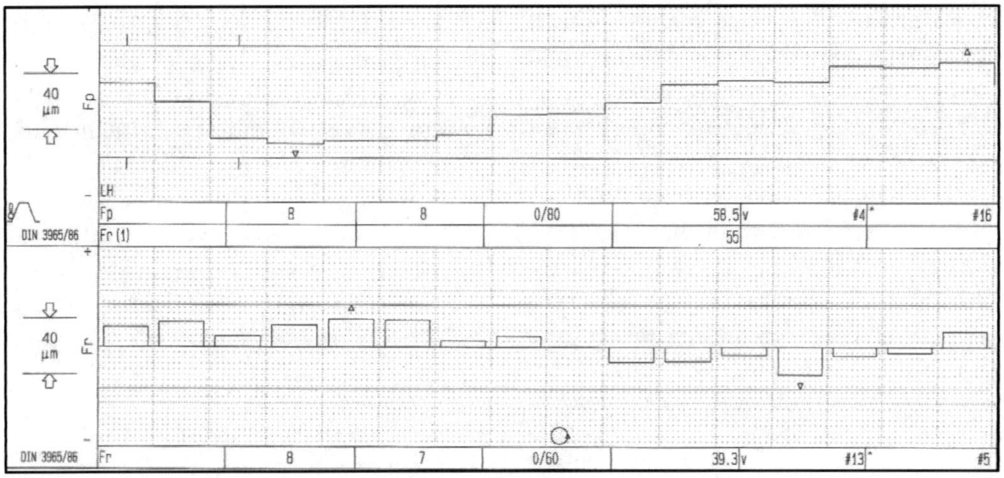

(a)

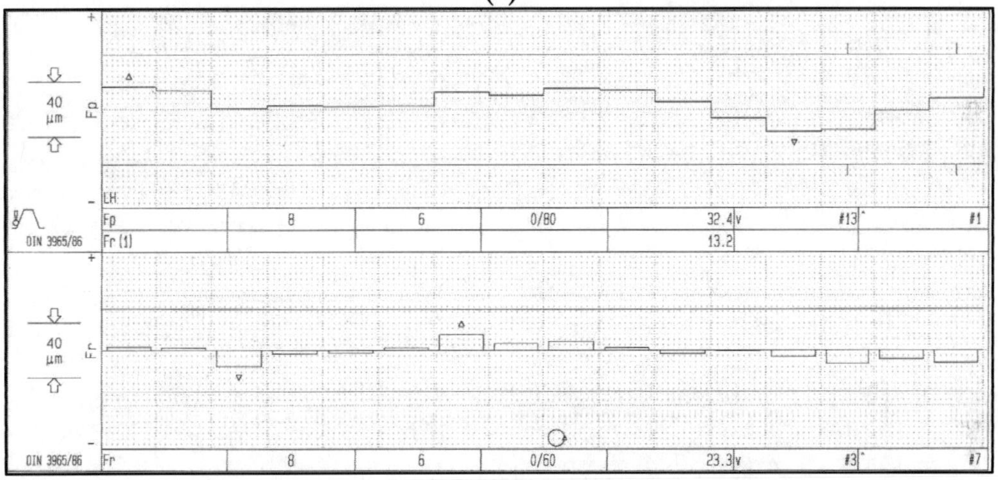

(b)

Fig. 4: Cumulative pitch error F_p' and total runout F_r' of straight bevel gear tooth for the best finished gear i.e. experiment no.3: **(a)** before PECH and **(b)** after PECH

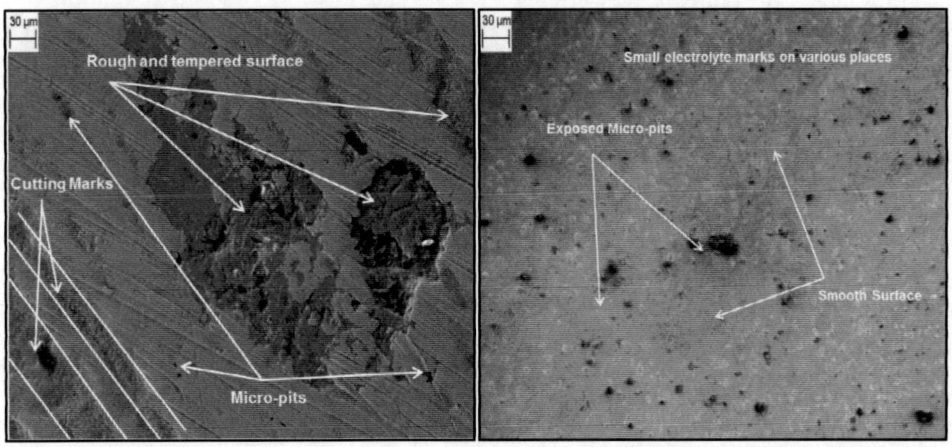

Fig. 5: SEM images of gear tooth flank surface at 400X magnification **(a)** unfinished gear; and **(b)** gear finished by PECH using the identified optimum input parameters.

This improvement in surface morphology is due to proper flushing of sludge products and controlled dissolution of the workpiece gear which restricts the implant of reactant products over workpiece surface. Vicker's hardness testing was done on a tooth flank surface of unfinished gear tooth flank and of a gear finished by PECH at the identified optimum input parameters. It was observed that the microhardness of the unfinished and finished gear tooth flanks is almost same i.e. in the range of 645-655 HV. From this result it can be concluded that PECH process does not affect the surface integrity of the finished gear.

4.0. Conclusions

Following conclusions can be drawn from the present study on the effects of electrolyte flow rate and rotary speed of the workpiece gear on the surface quality of PECH finished bevel gears:

- Electrolyte flow rate and rotary speed are found to be important parameters in improving the surface finish and microgeometry of the straight bevel gears by PECH. Lower electrolyte flow and an optimum value of workpiece rotary speed results in maximum improvement in surface finish and micro-geometry of the bevel gears. Their identified optimum values are 20 lpm and 40 rpm respectively.
- SEM images of the gear tooth flank finished by PECH revealed that the finished surface is free from mechanical and thermal distortions.
- These improvements in surface finish and microgeometry of the PECH finished bevel gears will help in improving their service life, running performance and may reduce the noise generation.
- It establishes that PECH as an economical and better alternative gear finishing process for simultaneous improving surface finish and microgeometry of the bevel gears.

5.0. References

1. Capello, G.; Bertoglio, S.; Annals of CIRP 1979, vol. 28, p. 103.
2. Wei; G.; Wang, Z.; Chen, C.; Precision Eng., 1987, vol. 9, p. 218.
3. He, F.; Zhang, W.; Nezu, K.; Japan Society of Mech. Eng. Int. J. (Series C) 2000, vol. 43, p. 486.
4. Naik, L.R.; Jain, N.K.; Sharma, A. K.; Proc. of 2nd Int. and 23rd AIMTDR Conf., 2008, p. 509.
5. Misra, J.P.; Jain, N.K.; Jain, P.K.; Proc. IMechE, Part B: J. Eng. Manuf., 2010, vol. 224, p. 1817.
6. Shaikh, J.H.; Jain, N.K.; Venkatesh, V.C.; Mater. Manuf. Process., 2013, vol. 28, p. 1117.
7. Shaikh, J.H.; Jain, N.K.; J. Mater. Process. Technol., 2014, vol. 214, p. 200.
8. Pathak, S.; Jain, N.K.; Palani, I.A.; Mater. Manuf. Process., 2014, vol. 29(11-12), p. 1461.
9. Pathak, S.; Jain, N.K.; Palani, I.A.; Mater. Manuf. Process., 2015, vol. 30(7), p. 836.

Microstructural, Mechanical and Tribological Properties of Electroless Duplex NiP/NiB Coatings on Steel

V. Vitry*, L. Bonin, F. Delaunois
Metallurgy Laboratory, Engineering Faculty, University of Mons, Mons, Belgium.
*Email: veronique.vitry@umons.ac.be

Abstract

Electroless nickel coatings present interesting mechanical, tribological and corrosion resistance properties. However, the two main kinds of electroless nickel coatings – Nickel-phosphorous (ENP) and nickel-boron (ENB) – are each more popular for some applications (corrosion resistance and mechanical applications respectively).

To assess the potential of cumulating the best features of each kind of coating, duplex electroless nickel coatings are investigated. The coatings used in the study were synthesized on mild steel substrates from a commercial bath in the case of nickel-phosphorous and from a lab-developed composition in the case of nickel-boron (with sodium borohydride as reducing agent and lead salts as stabilizer).
Single coatings of each electroless nickel composition with a thickness of 20 mμ were synthesized and compared with duplex layers (10 +10 μm) made as follow: ENP/ENB; ENB/ENP, ENP/ENP and ENB/ENB.

The morphology of all coatings was observed, their hardness, roughness and Taber abrasion index were measured and their resistance to scratch test was assessed. Systems with ENB on the top showed better abrasive wear performance, due to the higher hardness of nickel-boron and despite the higher roughness of ENB. The best resistance to scratch testing was obtained with single and double NiB coatings.

Surface Modification Technologies XXIX
Edited by T. S. Sudarshan and M. A. J. Somers

1.0. Introduction

Electroless nickel plating, based on the aqueous reduction of nickel salts by a chemical agent, has been used in the plating industry for more than 60 years[1,2]. This process knows several variations and is still investigated widely by researchers, notably because of its versatility due to the absence of an external current source[1,3].

Several reducing agents can be used to synthesize electroless nickel coatings, leading to a separation of the coatings in various classes: nickel-phosphorous, that is obtained with sodium hypophosphite (which is the most popular class); [1,2, 4–6]nickel-boron that is the product of the reduction by either sodium borohydride or amine-borane compounds;[1,3,7–11] and pure nickel (obtained with hydrazine)[1].

The properties of the various classes of electroless nickel are interesting and each class is known for being particularly suitable for some typical applications: nickel-phosphorous presents better corrosion resistance while nickel-boron is more useful in mechanical and wear applications[11,12]. The idea of combining both kind of coatings to improve the behavior of the whole system has thus appeared in research several years ago,[13-15] but none of the previous works did compare the duplex coatings with monolayers of similar thickness and with bilayers of a single material, which makes it impossible to assess which of the properties are due to the presence of two layers and which to the duplex nature of the coating. In this paper, mono and bi-layers of both nickel-boron (ENB) and nickel-phosphorous (ENP) are compared to coatings in which both materials are combined (duplex coatings).

2.0. Methods

2.1. Materials

Electroless nickel mono and bi-layer, as well as duplex coatings were deposited on 10*10 cm plates of St-37 mild steel with a thickness of 1mm, some of which had a central hole of 7mm diameter to allow Taber abrasion testing. All samples were ground with with SiC paper up to 1200 mesh, cleaned and degreased with acetone and etched 2 to 2.5 minutes in 30 vol. % HCl immediately before immersion in the electroless plating bath.

Two baths were used for this study: a commercial bath for ENP (Niklad 767 from McDermid), leading to a medium phosphorous coating (approximately 7 wt.% P) and our own bath, that produces a coating with 5 wt.% B and is stabilized by lead salts, for ENB [7,8,16].

The plating time needed to attain layers of 10 and 20μm was determined by the use of a calibration curve obtained from previous experiments by the weight gain method. The accepted error on plating was +/- 2 μm per layer. The following coatings were made: ENB and ENP monolayers with a thickness of 20 μm; ENB and ENP bi-layers of 20μm (10μm +10μm); and duplex coatings: 10μm ENB + 10μm ENP and 10μm ENP + 10μm ENB.

2.2. Analytical methods

The morphology of samples was observed with a Hirox digital microscope on the surface and a Jeol JSM 5900 LV scanning electron microscope on polished cross sections. The structure of the deposits was investigated by X-ray diffraction (XRD), with a Siemens D500 X-rays θ–2θ apparatus applying Co $K\alpha$ (1,79 Å). The roughness of all samples was measured by the mechanical stylus method with a Zeiss 119 Surfcom 1400D-3DF apparatus. Hardness of the various systems was measured on the surface by Vickers microindentation with a load of 100gf and a holding time of 20s. The Taber abrasion test was used to characterize the wear resistance of the coatings, the tests was carried out on a 5155 Taber apparatus with CS-17 abrasive rubber wheels, under a load of 1000g and with a rotating speed of 72 rpm.

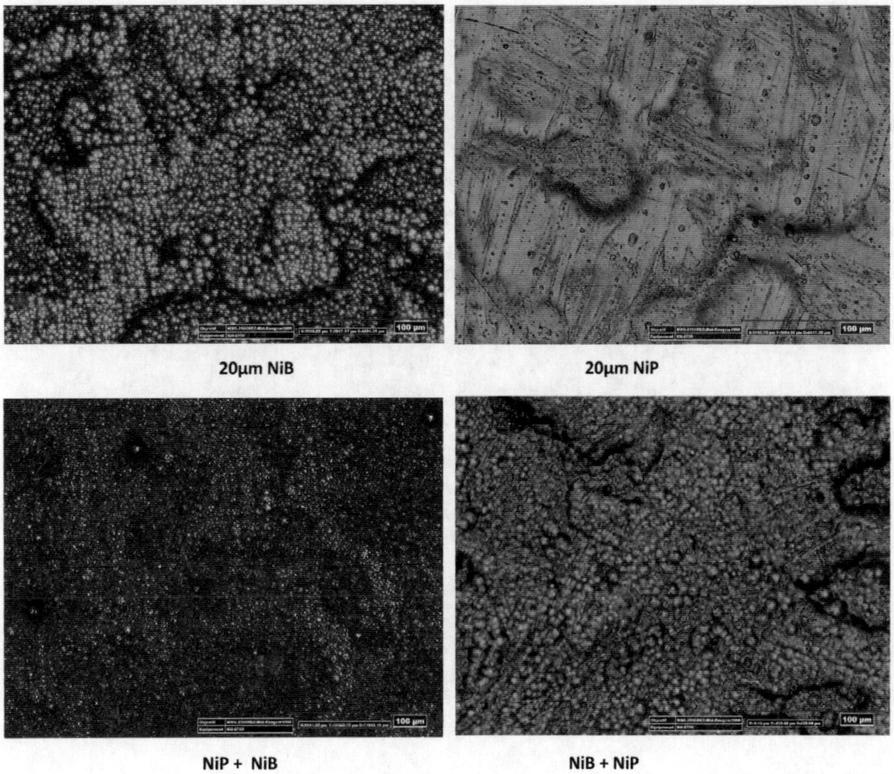

Fig. 1: Digital microscopy images of surface morphology of ENB, ENP and Duplex coatings

The scratch resistance was measured by the continuous increase method, with a CSEM scratch tester and the following conditions: load rate - 100N.min-1; advance rate - 6.75 mm.min-1; final load - 150 N; scratch length - 10 mm; tip - diamond Rockwell stylus with a radius of 200 µm. The resulting scratches were observed with a Hirox digital microscope.

3.0 Results

3.1 Morphology and structure

The surface morphology of the coated systems is presented on Fig. 1. The use of a bilayer of a single material does not modify this morphology compared to a monolayer and is thus not shown. The monolayer coatings present typical surface morphologies for their type of coating: cauliflower-like texture for ENB and planar for ENP. The morphology of duplex coatings is something in between. The cross sections observed by SEM are shown on Fig. 2. All coated systems present a thickness in the desired range (~20 µm) and the distinct layers can be easily observed for the bilayer and duplex systems, as well as the morphology of the nickel-boron layers. Nickel-phosphorous however, cannot be etched by nital and stays thus featureless except for the delimitation of the layers.

Fig. 3 presents the X-ray diffraction graphs for all the systems. All coatings present only one peak, at 52°, which corresponds to supersaturated nickel, either nanocrystalline or amorphous. However, the peaks related to samples containing only ENP or where ENP is on top of the system are significantly sharper than the others, which suggests that ENP is more crystallized than ENB.

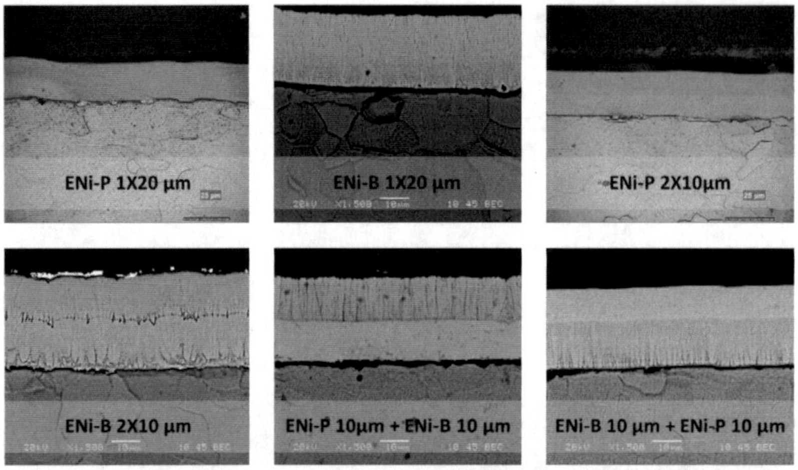

Fig. 2: SEM cross section observation of mono and bilayers of ENP and ENB and of Duplex coatings

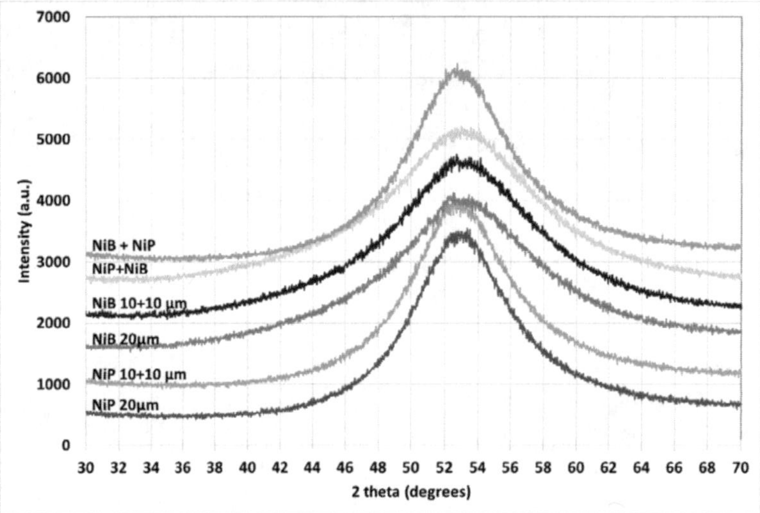

Fig. 3: X-ray diffraction graph of mono and bilayers of ENP and ENB and of Duplex coatings

3.2 Roughness, hardness and wear resistance

Roughness values are shown on Table 1. They are on average very low and do not present important standard deviation but consistent differences are observed between the systems: the roughness of systems where ENB is the topmost layer is higher and bilayer systems are rougher than monolayers of the same kind. However, using a duplex coating does not increase the roughness compared to a bilayer system.

The surface hardness of the samples is shown on Table 2. All the systems containing ENB are much harder than those that only contain ENP, even the duplex ones that contain equivalent amounts of ENP and ENB. The duplex system where ENB is the top most layer is nearly as hard as mono and bilayer ENB while that with ENP on top is closer to the harness of ENB than that of ENP.

Taber Wear index for all coatings is also shown in Table 2. The coatings that do not contain ENB present the worst TWI of all.

Table 1: Roughness, hardness and Taber Wear Index (TWI) of electrolesss mono and bi-layers and of duplex systems

	Ra (μm)	**Rq (μm)**	**Rz (μm)**	**Rp (μm)**	**Rv (μm)**	**Rt (μm)**
ENP 20μm	0.093+/- 0.008	0.115+/- 0.009	0.418+/- 0.021	0.208+/- 0.013	0.209+/- 0.008	1.407+/- 0.096
ENP 10 + 10μm	0.123+/- 0.003	0.15+/- 0.003	0.523+/- 0.006	0.259+/- 0.006	0.264+/- 0.003	2.173+/- 0.246
ENB 20μm	0.162+/- 0.011	0.202+/- 0.012	0.782+/- 0.045	0.366+/- 0.017	0.416+/- 0.027	2.293+/- 0.152
ENB 10 + 10μm	0.198+/- 0.011	0.244+/- 0.013	0.917+/- 0.041	0.432+/- 0.019	0.485+/- 0.023	3.747+/- 0.411
ENP + ENB	0.119+/- 0.001	0.149+/- 0.001	0.601+/- 0.002	0.287+/- 0.002	0.314+/- 0.003	1.747+/- 0.034
ENB + ENP	0.106+/- 0.01	0.133+/- 0.011	0.509+/- 0.038	0.247+/- 0.018	0.262+/- 0.021	2.26+/- 0.567

Table 2: Hardness, Taber Wear Index (TWI) and critical load for first damage of electrolesss mono and bi-layers and of duplex systems

	Hardness (hv_{100})	**TWI**	**Lc (N)**
ENP 20μm	551.35 +/- 2.1	35.9+/-0.4	15
ENP 10 + 10μm	549.31 +/- 6.2	34.9+/-2.4	15
ENB 20μm	872.09 +/- 5.5	31.1+/-1.1	35
ENB 10 + 10μm	875.5 +/- 4.6	27.6+/-2.3	25
ENP + ENB	860.66 +/- 6.1	27.2+/-0.7	15
ENB + ENP	738.9 +/- 6.1	28.9+/-1.8	10

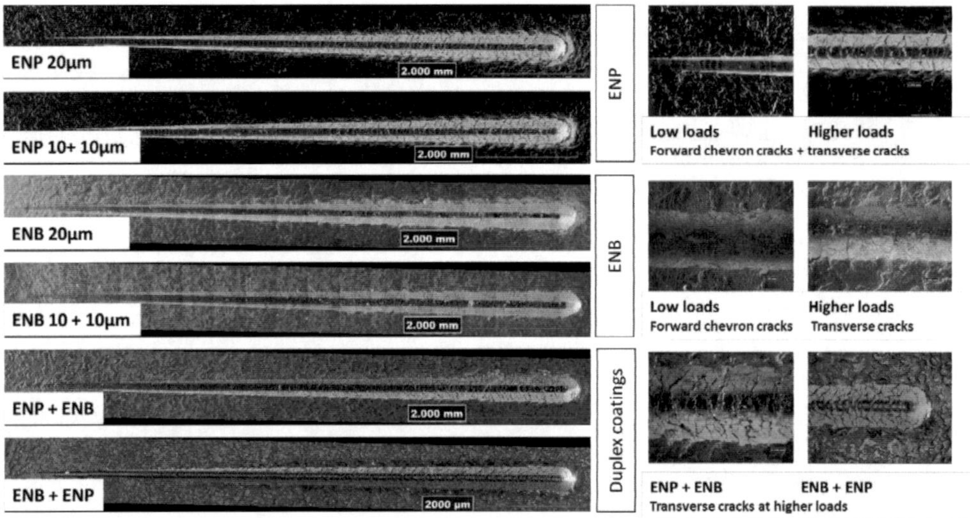

Fig. 4: Digital microscopy observation of scratch tests on mono and bilayers of ENP and ENB and of Duplex coatings. Global observation of scratches and details of the typical damage

It appears that the use of bilayers and duplex coatings improves the abrasive wear resistance of coatings: bilayers present a significantly lower TWI than their monolayer counterparts and duplex coatings are more resistant to abrasive wear than the bilayer of their topmost material.

3.3. Scratch tests

All coated samples were submitted to scratch testing. Global view of the scratches and details of the typical damage for the various types of coatings are shown on Fig. 4. The critical load for first damage for all coatings, obtained by a combination of acoustic emission and microscopy observation, is given in Table 2.

None of the coatings presented complete failure or scaling. Mono and bilayers of a same material present similar damage and critical loads in the same range but the use of bilayers seems to decrease the critical load for first damage. As expected from previous results[12], ENB shows a better scratch tests behavior than ENP.

Duplex coatings have critical loads for first damage smaller or equal to that of ENP systems and the damage they present at the end of the scratch is similar to that observed in those systems.

4.0. Discussions

The use of Duplex systems modifies the features of electroless coatings in a way that they acquire behavior that generally in between that of ENP and ENB. The surface morphology of those coatings, is relatively similar to that of ENB, even if that layer is at the bottom. This is due to the very good ability of ENP to conserve surface shape: the cauliflower-like texture of ENB is still observed after 10 µm of ENP have been put on top of the previous ENB coating. This is confirmed by the roughness of the coatings: ENP is smoother than ENB when used in similar configurations.

The roughness of the system is lowered by the use of duplex coatings, compared to single material bilayers. This was expected in the case of ENB but not of ENP. The exact causes of this phenomenon still have to be determined. The increase of roughness between single material mono and bi-layer may play a role in the observed increase of wear resistance for bilayers compared with monolayers that have the same chemistry (and also the same hardness). However, other effects due to the presence of two layers may not be ruled out at the present time. Further investigation will be carried out to ascertain this.

The values of hardness and Taber Wear Index measured on the samples are as expected:[12] the ENB mono- and bilayers present much higher hardness and wear resistance than the ENP. The Duplex systems are somewhere between the others, with better properties for the system where ENB is the top layer. This is not surprising because the top layer is the most solicited in hardness and abrasive wear testing. For scratch tests resistance, better behavior was also observed for the ENB coatings. However, in this case, systems with two separate layers behave less effectively than the monolayers. This may be due to the presence of a new interface in the coating. The Duplex coatings have thus a double decrease of their properties: they contain ENP (which is known to have a lesser scratch resistance than nickel-boron) and are constituted of two layers (and thus present one more interface than monolayers). They present thus the worst results of all the systems, even if their scratch resistance stays relatively good.

5.0. Conclusion

In this paper, we investigated several aspects of mono and bilayer electroless nickel coatings, as well as duplex coatings. Here are the main observations:

- The roughness of electroless nickel monolayers, bilayers and duplex coatings is low (below 0.2 μm) but the smoothest coatings are those that do not contain ENB. The highest roughness was observed on the bilayer ENB (10 +10 μm).
- Mono and bilayer ENB coatings are harder, have a better wear resistance and better scratch test behavior that similar ENP coatings, as expected from literature.
- The replacement of a 20μm monolayer of ENP or ENB by two 10μm layers of the same material does not change the overall hardness of the system. However, roughness and the abrasive wear resistance are increased and the scratch resistance is decreased, probably due to the presence of a new interface.
- The roughness of duplex EN bilayers is quite is in the same range as that of monolayer coatings and lower that bilayer coatings.
- Duplex EN coatings present a better abrasive wear resistance that coatings made of their top-layer material. The best wear resistance is thus obtained by ENP/ENB coatings. That of ENB/ENP is lower than the previous one and than that of ENB 10 + 10 μm, but it is higher than ENB 20μm and also than any ENP coating.
- Due to the fact that the presence of ENP and the creation of a new interface in the coated system are detrimental factors for scratch resistance, Duplex coatings present the worse scratch behavior of all the systems studied in this paper.

6.0. Acknowledgments

The authors wish to thank Mr E. Juste from BCRC for the realization of the scratch tests. They also wish to thank MM S. Gobeaut and F. Tosar for their implication in experimental work.

7.0 References

1. Riedel, W.; Electroless Nickel Plating. Finishing Publication LTD., London, 1991.
2. Schlesinger, M.; Paunovic, M.; Modern Electroplating: Fifth Edition. 2011.

3. Vitry, V.; Delaunois, F.; Anti-Abrasive Nanocoatings. Elsevier, 2015.
4. Agarwala, R.C.; Agarwala, V.; Electroless alloy / composite coatings : A review, Sadhana, 2003, p. 475.
5. Sahoo, P.; Das, S.K.; Tribology of electroless nickel coatings - A review, Mater. Des., 2011, vol. 32(4), p. 1760.
6. Brenner, A.; Riddell, G.E.; Nickel plating on steel by chemical reduction, J. Res. Natl. Bur. Stand. 1946., vol. 37, p. 31.
7. Vitry, V.; Electroless Nickel-Boron deposits: Synthesis, formation and characterization; Effect of heat treatments; Analytical modeling of the structural state, University of Mons, Belgium, 2010.
8. Delaunois, F.; Petitjean, J.P.; Jacob-Dulière, M.; Liénard, P.; Surf. Coatings Technol., 2000, vol. 124, p. 201.
9. Ogihara, H.; Udagawa, K.; Saji, T.; Surf. Coatings Technol., 2012, vol. 206(11-12), p. 2933.
10. Lacourcelle, L.; Nickelage chimique, in Techniques de l'Ingénieur, 1995.
11. Sudagar, J.; Lian, J.; Sha, W.; J. Alloys Compd., 2013, vol. 571, p. 183.
12. Vitry, V.; Sens, A.; Delaunois, F.; Mater. Sci. Forum, 2014, vol. 783-786, p. 1405.
13. Vitry, V.; Sens, A.; Kanta, A.F.; Delaunois, F.; Surf. Coatings Technol., 2012, vol. 206(16), p. 3421.
14. Delaunois, F.; Lienard, P.; Surf. Coatings Technol., 2002, vol. 160, p. 139.
15. Narayanan, T.S.N.S.; Krishnaveni, K.; Seshadri, S.K.; Mater. Chem. Phys., 2003, vol. 82, p. 771.
16. Vitry, V.; Kanta, A.F.; Delaunois, F.; Ind. Eng. Chem. Res., 2012, vol. 51(1), p. 9227.

Response of Nanometric WC-12Co Microwave Clads to Abrasive Wear

S. Zafar, A.K. Sharma*
Mechanical and Industrial Engineering Department,
Indian Institute of Technology Roorkee, Roorkee, India- 247667.
*Email: akshafme@gmail.com, akshafme@iitr.ac.in

Abstract

In the present work abrasive wear performance of the micrometric (MM) and nanometric (NM) WC-12Co clads deposited through microwave hybrid heating has been studied. The microstructures of the MM and NM WC-12Co microwave clads were studied through SEM. It was found that carbide phases are more uniformly distributed in NM clads rather than classical skeleton-type structures in the MM clads. Phase analysis of the microwave clads was carried out using X-ray diffraction. The abrasive wear characterstics of the WC-12Co clads were investigated using dry sand abrasive wear test as per ASTM G65 with silica sand as abrasive at different speeds. The NM microwave clads exhibited enhanced wear resistance at higher speeds while compared to the MM clads. This can be attributed to higher volume fraction of the nanostructured carbides present in the clad layer. Mechanism of material dislodgement during abrasion was studied using SEM and are discussed through microstructural evidences.

1.0. Introduction

Components deployed in many industries such as construction and agricultural equipments, mining etc. are prone to abrasive wear. The WC-Co based materials are known for their exceptional abrasive wear resistance[1–3]. Recent work has shown that abrasive wear performance of cermets depends on size and distribution of the carbide phase, hardness of the carbide phase relative to hardness of the abrasive and volume fraction of the carbide phase[4–6]. Hence nanostructured WC-Co based materials offer improved mechanical, tribological and chemical properties that their micrometric counterparts. Thus, in order to reduce abrasion-induced failure of the mechanical components, use of such materials could be a pragmatic solution. Thermal spray based methods have been a popular choice to deposit WC based coatings; however lack metallurgical bonding, chemical inhomogeneity and porosity are some of the limitations of the process.[7] Moreover, decarburisation of the carbide phase during thermal spray results in reduced wear performance of coatings[8,9].

On the other hand, microwave cladding has been an emerging surface modification technique owing to development of uniform microstructures, absence of interfacial cracking, less thermal distortion[10–13]. It was earlier reported that volumetric heating characterstics of microwave reduced decomposition of the carbide phase in WC-Co based clads[14–16]. Moreover, nanostructured materials posses higher surface area to volume ratio than their conventional counterparts, which results in better microwave absorption and hence uniform and volumetric heating of the clad powder. However, performance of these clads are yet to be assessed while subjected to different test environments. Accordingly, in the present work microwave clads of MM and NM WC-12Co powders were developed on SS-304 substrates and the clads were evaluated using SEM and XRD. Further, response of the MM and NM clads to abrasive wear was assessed using dry sand rubber wheel test under different speeds. The worn surfaces of the WC-12Co clads were studied using SEM to identify the wear mechanisms.

2.0. Experimental Procedure

2.1. Material details and cladding process

Commercially available micrometric (MM) and nanometric (NM) WC-12Co powders were used in the present study. Data concerning powder morphology, composition and particle size are shown in Table 1. An energy dispersive X-ray spectrometer (EDS) was used to evaluate the composition of the clad powders. The morphology of the MM and NM clad powders was studied using SEM and is presented in Fig. 1. The MM powder exhibits a spherical agglomerated morphology while the NM powder illustrates an irregular crushed morphology. An austenitic stainless steel (SS-304) was used as the substrate material in the present investigation. Samples from SS-304 plates were machined to 25 × 25 × 8 mm^3 size and were polished to attain an initial surface roughness (R_a) of 0.15 μm.

The clad powder was manually placed over the substrate maintaining an approximate thickness of 1 mm. The WC-12Co clads were developed in an industrial microwave applicator (Make: Enerzi Microwave Systems, Model: MH-1514-101-V6). The substrate with preplaced powder was placed inside the applicator and was then subsequently exposed to microwave radiation. An alumina glass wool casket was used to envelop the hybrid heating arrangement placed at the centre of the microwave cavity. This casket allows microwaves to pass through and insulates the heat generated. Charcoal powder was used as the susceptor material.

Fig. 1: SEM micrographs of **(a)** micrometric and **(b)** nanometric WC-12Co clad powders

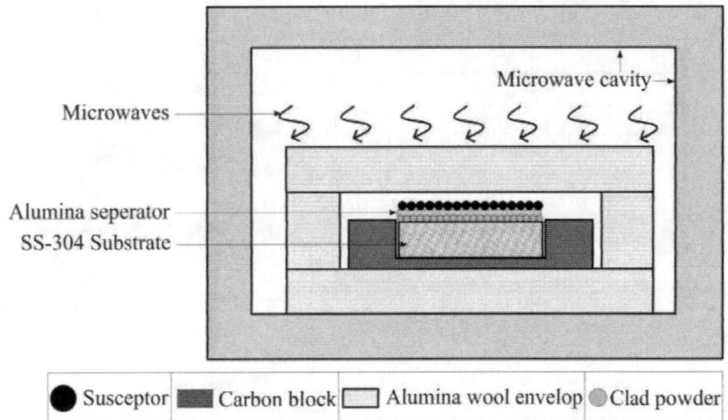

Fig. 2: Schematic of the MHH experimental setup used to develop the WC-12Co clads

Table 1: Characterstics of the clad powders

Clad powder	Composition	Average particle size
Micrometric (MM)	77.33 W, 12.45 Co, 10.22 C	40- 45 μm
Nanometric (NM)	74.63 W, 12.52 Co, 12.85 C	100-200 nm

Charcoal powder not only provides hybrid heating as well as reduces the thermal gradient which facilitates formation of crack free clads. To avoid any possible contamination of clad powder by the susceptor, 1 mm thick alumina plate (purity 99%) was used as a separator. The metallic substrate was selectively masked using a carbon block to avoid reflections of microwaves as illustrated in Fig. 2. A detailed description of mechanism of clad formation through MHH has been detailed elsewhere[10,12,15,16]. Optimised cladding parameters for MM and NM powders are presented in Table 2.

2.2. Clad characterisation and abrasive wear testing of the clads

A low speed diamond saw (Make: Chennai Metco, Model: BAINCUT LSS) was used to section clad specimens along the thickness. The sectioned specimens were then polished using standard metallographic techniques to attain mirror finish. The microstructures of the MM and NM clad layer were analysed using a SEM (Make: Carl Ziess, Model: Gemini Ultra Plus). The phases present in the clad layer were analysed using a XRD (Make: Bruker AXS, Model: D8) using Cu-Kα X-ray within the scan range of 30° to 80° at a scan rate of 1° min $^{-1}$.

Table 2: Microwave cladding parameters for developing the MM and NM WC-12Co clads

Parameters	**Values**	
Powder	MM	NM
Powder preheat temperature (°C)	120	120
Power (kW)	1.4	1.1
Frequency (GHz)	2.45	2.45
Time of exposure (s)	600	540
Processing environment	Ambient	

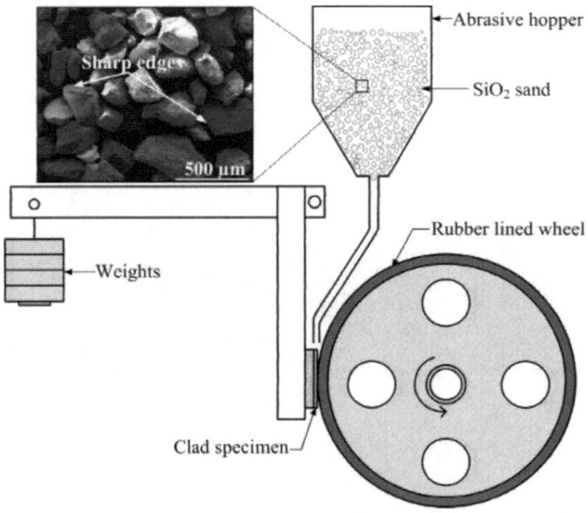

Fig. 3: Schematic diagram of dry sand rubber wheel test setup. Inset: SEM micrograph of the sand

A dry sand rubber wheel test was used to evaluate the abrasive wear response of the MM and NM microwave clads (Fig. 3). Clad specimens (size: 25 × 25 × 8 mm^3) were turned against rubber wheel under a load of 60 N, sand flow rate of 80 g min^{-1} as illustrated in Fig. 3. The test was carried out at three speeds of 1.5, 2.0 and 2.5 m s^{-1}. Sand (SiO_2) particles, having an average size of 300 µm, were used as the abradant.

The sand particles exhibit angular and blocky morphology as observed from the inset SEM micrograph in Fig. 3. The specimens were ultrasonically cleaned before and after the test. A microelectronic weight balance (Make: Mettler Toledo, Model: ML 104) was used to weigh the specimens with an accuracy of 0.1 mg. First 300 s were not taken in account in order to attain a steady state. The weight loss experienced by the specimens was measured after every 300 s and was used to calculate the cumulative weight loss. Further, the worn surfaces of the MM and NM clads were studied using SEM.

3.0. Results and Discussion

3.1. Clad microstructure

Typical SEM micrographs illustrating microstructures of the WC-12Co MM and NM microwave clads are shown in Fig. 4. The WC-12Co MM and NM clads of thickness of 1 mm (approximately) were developed using MHH technique. The developed clads exhibited a well developed metallurgical bonding without any visible interfacial cracks. The MM clad layer consists of white skeleton structured reinforcing carbides in the clad layer (Fig. 4(a)).

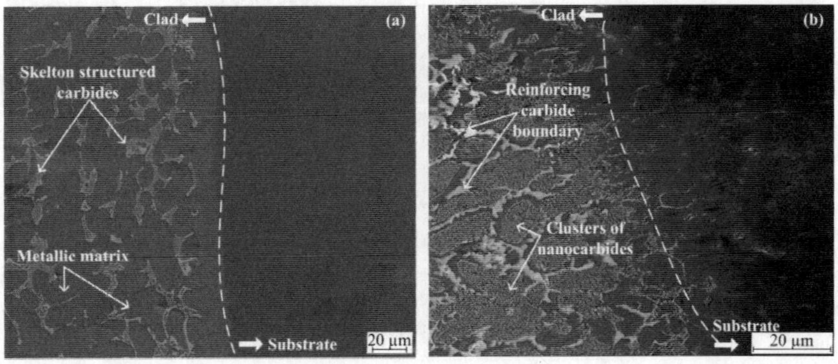

Fig. 4: A typical SEM micrograph of transverse section of WC-12Co **(a)** MM and **(b)** NM microwave clad

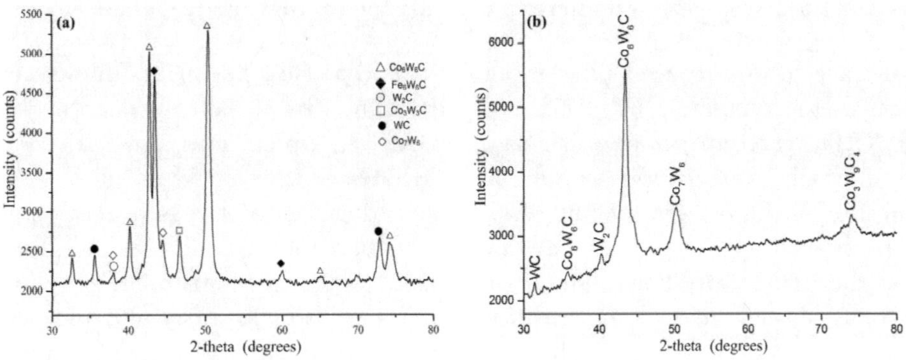

Fig. 5: XRD spectra of WC-12Co **(a)** MM and **(b)** NM microwave clads

These tungsten based carbides were observed to be uniformly distributed in the Co based matrix that provide reinforcement to the clad layer.[16] The microstructure of the NM microwave clad exhibits presence of clusters of nanocarbides in an interconnected WC based network (Fig 4(b)). However, depleted of carbides could be identified in the region close to the network, possibly due to diffusion. Enhanced reinforcement in the clad layer could result in higher wear resistance of the NM clads. Further, the average volume fraction of the carbides in the MM and NM microwave clads was evaluated using an image analysis tool (ImageJ, Version: 1.46r). The image analysis was carried out at five different locations in the clad layer. The NM WC-12Co exhibited approximately 48% higher carbide volume fraction as compared to the MM clads. Higher volume carbide volume fraction also indicates towards higher hardness of the NM clad layer as compared to the MM clad.

The XRD spectrum of the WC-12Co MM and NM microwave clads is shown in Fig. 5 (a) and (b) respectively. It is clear from the XRD patterns that the decomposition of the WC phase was suppressed attributed to the volumetric heating characterstics of MHH. This can be confirmed by the presence of trace amount of W_2C phase. Dominant presence of Co_6W_6C in the MM and NM clads can be evidenced from Fig. 5. The Co_6W_6C intermetallic exhibits properties like bulk modulus of the order of 462 GPa which is higher than that of WC (approximately, 421 GPa)[17]. Hence, the presence of Co_6W_6C indicates higher toughness of the reinforcing carbides in the microwave developed clads.

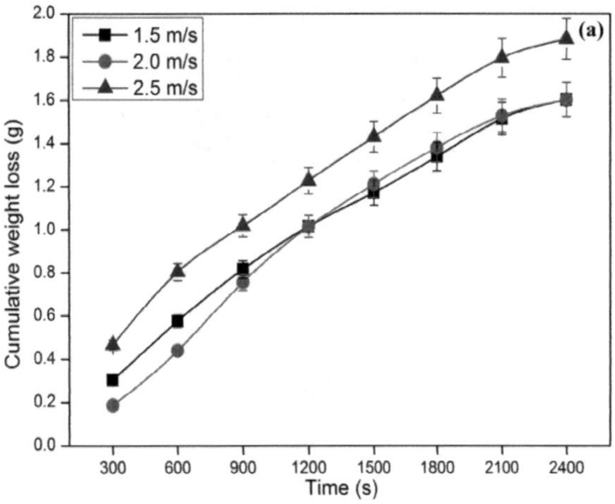

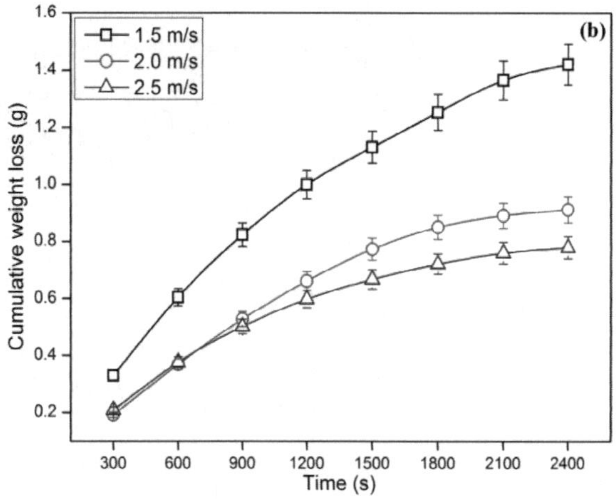

Fig. 6: Cumulative weight loss characterstics of WC-12Co **(a)** MM and **(b)** NM microwave clad

Presence of traces of amorphous phases can be evidence at approximately 62°, 66° and 70° in the XRD spectrum of the MM clad. However no such observation could be made from the XRD pattern of the WC-12Co NM microwave clad. It can be confirmed that the presence of non crystalline phases was reduced in the NM clads, that could lead to enhanced abrasive wear resistance.

3.2. Abrasive wear

The cumulative weight loss characterstics of the WC-12Co MM and NM microwave clads are shown in Fig. 6. It can be observed that under various speeds the NM clads exhibit low cumulative weight loss as compared to the MM clads. However, it interesting to observe that the trend of cumulative weight loss has been opposite in case of the MM and NM clads. The material loss from the MM clad increases as the speed increases, while for the NM clad the material loss decreases as the speed increases. It can be observed from Figs. 7 (a-c), that the worn surfaces of the MM clad exhibits deeper abrasion marks with increasing speed, attributed to the removal of the binder. As the speed increases undermining of the carbide phase takes places due to removal of the binder by the abrasive action of the sand particles, which leads to pull out of the carbide as observed from Fig. 7(c). However, worn surfaces of the NM clad exhibits mild abrasion, pitting and microcutting as the main wear mechanisms (Figs. 7(d-f)) The presence of clusters of nanocarbides, as observed in the clad microstructure (section 3.3), provide better reinforcement and minimise the exposure of the relatively soft matrix to sand particles. But at higher speeds some cavities do occur on the worn surface of the NM clad (Fig. 7(f)). For given cobalt content, as the carbide grain size decreases the mean free path of the matrix is reduced that result in greater constraint and reduced tendency for binder phase to get removed due to the abrasive action of the sand particles[9]. High carbide volume fraction in the NM clad reduces the mean free path of the binder. Also higher number of grain boundaries in the NM clad impedes slip, which leads to restricted crack propagation. Therefore, it can be substantiated that the NM clad exhibit better abrasive wear resistance attributed to higher carbide volume fraction and homogenous structure in the clad layer.

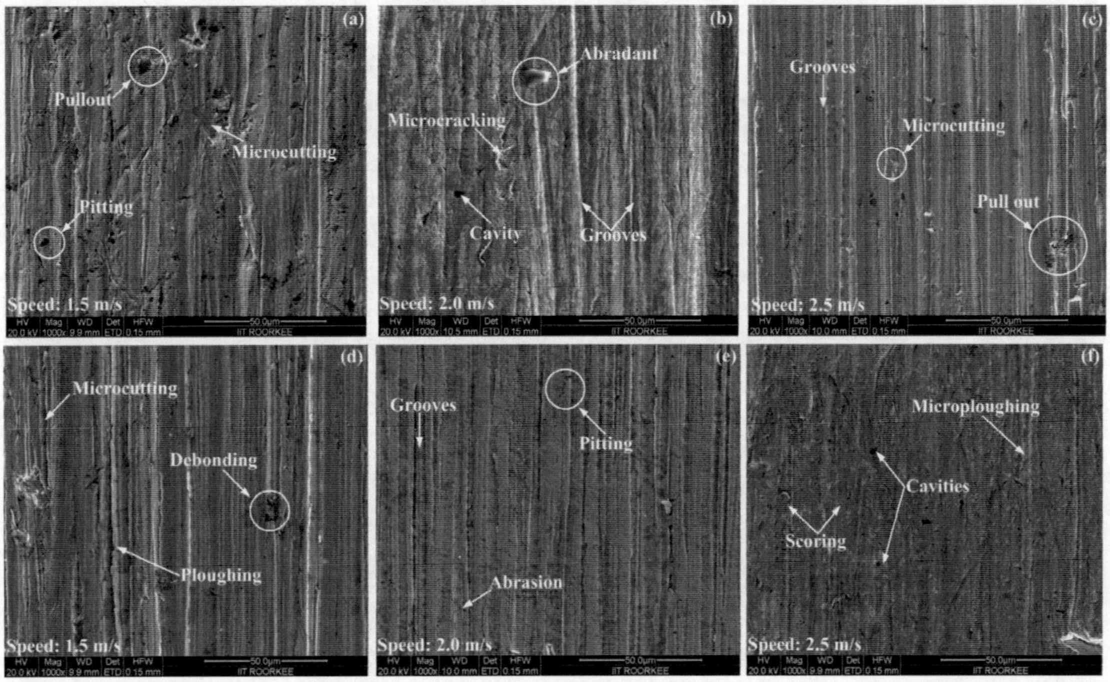

Fig. 7: Typical SEM micrographs of worn surfaces at various speeds **(a-c)** MM clads and **(d-f)** NM clads.

4.0. Conclusions

In the present work, WC-12Co clads were deposited using MM and NM powders on SS-304 substrate using MHH technique. The MM clads consists of skeleton structured carbides distributed in the Co based matrix, while the NM WC-12Co clads consist of uniformly distributed clusters of nanocarbides. Presence of amorphous phases in the NM clad was reduced due to volumetric heating during microwave irradiation. The abrasive wear response of the clads reveals that NM clads exhibit higher abrasive wear resistance at higher speed due to the presence of clusters of nanocarbides in the NM clad. In the MM clads, on the other hand, the unsupported carbides gets easily fractured and/or pulled out leading to increased material loss owing to higher interaction area of the relatively soft matrix.

5.0 References

1. Gant, A.J.; Gee, M.G.; Roebuck, B.; Wear, 2005, vol. 258, p. 178.
2. Wang, H.; Song, X.; Wei, C.; Gao, Y.; Guo, G.; J. Mater. Sci. Technol., 2013, vol. 29, p. 1067.
3. Afzal, M.; Ajmal, M.; Khan, A.N.; Tribol. Transac., 2014, vol. 57, p. 94.
4. Qiao, Y.; Fischer, T.E.; Dent, A.; Surf. Coat. Technol., 2003, vol. 172, p. 24.
5. Dent, A.H.; Depalo, S.; Sampath, S.; J. Therm. Spray Technol., 2002, vol. 11, p. 551.
6. Hazra, S.; Sabiruddin, K.; Bandyopadhyay, P.P.; Surf. Eng., 2012, vol. 28, p. 37.
7. Afzal, M.; Ajmal, M.; Khan, A.N.; Hussain, A.; Akhter, R.; Optics and Laser Technol., 2014, vol. 56, p. 202.
8. Karimi, A.; Verdon, C.; Barbezat, G.; Surf. Coat. Technol., 1993, vol. 57, p. 81.
9. Wang, Q.; Chen, Z.H.; Ding, Z.X.; Tribol. Int., 2009, vol. 42, p. 1046.

10. Gupta, D.; Sharma, A.K.; Surf. Coat. Technol., 2011, vol. 205, p. 5147.
11. Singh, S.; Gupta, D.; Jain, V.; Sharma, A.K.; Mater. Manuf. Process., 2014, vol. 30, p. 1.
12. Gupta, D.; Sharma, A.K.; J. Manuf. Process., 2014, vol. 16, p. 176.
13. Zafar, S.; Bansal, A.; Sharma, A.K.; Arora, N.; Ramesh, C.S.; Surf. Eng., 2014, vol. 30, p. 852.
14. Gupta, D.; Sharma, A.K.; Wear, 2011, vol. 271, p. 1642.
15. Gupta, D.; Sharma, A.K.; J. Mater. Eng. Perf., 2012, vol. 21, p. 2165.
16. Zafar, S.; Sharma, A.K.; Mater. Charac., 2014, vol. 96, p. 241.
17. Suetin, D.V.; Shein, I.R.; Ivanovskii, A.L.; Physica B: Physics of Condensed Matter, 2009, vol. 404, p. 3544.

Formation of Protective Composite Coatings on Mg Alloy by Plasma Electrolytic Oxidation Method with Using Organofluorine Materials

S.V. Gnedenkov, S.L. Sinebryukhov,
Institute of Chemistry FEB RAS,
159, pr. 100-letiya Vladivostoka, Vladivostok 690022, Russia.

D.V. Mashtalyar, K.V. Nadaraia*,
Institute of Chemistry FEB RAS,
159, pr. 100-letiya Vladivostoka, Vladivostok 690022, Russia.

Far Eastern Federal University,
8, Sukhanova St., Vladivostok 690950, Russia.

A.N. Minaev
Far Eastern Federal University,
8, Sukhanova St., Vladivostok 690950, Russia.
*Email:nadaraiakv@mail.ru

Abstract

The results of surface modification of magnesium alloys by plasma electrolytic oxidation (PEO) and subsequent treatment in suspension of superdispersed polytetrafluoroethylene (SPTFE) and telomeric solution of tetrafluoroethylene (TFE) have been presented. Composite coatings were obtained by dipping with subsequent heat treatment. Electrochemical, tribological properties and wetting ability of obtained protective composite coatings were investigated. The best protective properties were obtained for coatings with fourfold application in SPTFE suspension. The obtained coatings decrease the corrosion current ($5.4\ 10^{-11}$ A cm^{-2}) and wear ($7.6\ 10^{-7}$ mm^3 $(N\ m)^{-1}$), increase polarization resistance ($4.9\ 10^8$ Ω cm^2) and impedance modulus ($1.9\ 10^9$ Ω cm^2) by orders of magnitude compared to unprotected magnesium alloy and base PEO-coating. The highest value of contact angle (CA) was obtained for coatings with triple application in telomeric solution. CA for such composite coatings attaint to 168±2°.

1.0. Introduction

Today magnesium alloys are increasingly being used in industries such as aviation, automotive, producing of high-tech devices and in their components. But low corrosion and wear resistance substantially restrict the application scope of this material. One of the possible ways to protect magnesium alloys from the negative impact is the formation coatings by plasma electrolytic oxidation (PEO)[1]. PEO-method allows forming on the magnesium alloy protective layers, which have a convolute, rough surface. Such structure can be serve as a basis for application on the PEO-surface polymeric materials in order to provide additional protection for this product[2]. Particular interest this regard are fluoropolymers due to their chemical stability. This paper presents the results of investigation protective properties of composite coatings (CC) on magnesium alloys with various organofluorine materials.

2.0. Methods

Rectangular plates of magnesium alloy MA8 (system Mg-Mn-Ce) were used as samples. Silicate-fluoride solution was used as electrolyte for PEO. The process of coating formation was realized using the tyristor rectifier as a power supply. Coatings were formed in bipolar PEO mode. In the first step the anodic current density was specified of 0.61 A cm^{-2} (under incresing voltage up to 270 V), cathodic component was kept potentiostatically at –40 V. In the second step voltage was gradually decreased down to 200 V in the anodic part and down to –10 V in the cathodic. The duration of the first and second phases of the PEO was 200 s and 600 s, respectively.

Solution of telomers of TFE (trademark Cherflon®)[3] and aqueus suspension of SPTFE (trademark Forum®)[4] were used as fluoropolymer materials for formation of CC. The polymer was deposited by immersing of the sample with PEO-coating in different suspensions with subsequent heat treatment. Time of the immersion in suspensions was not more than 10^{-15} s. After application the samples were air dried for 20 minutes at 25°C. Next step was heat treatment at 250°C for coatings, which were obtained by application of telomeric solution (the number of applications was 3) (hereinafter CC-1) and at 315°C for coatings, which were obtained by application of SPTFE (the number of applications was 4) (hereinafter CC-2).

Electrochemical properties of coatings were investigated using electrochemical system VersaSTAT MC (Princeton Applied Research, USA). Measurements were made in a three-electrode cell in a 3 % NaCl solution at room temperature. Platinum gauze was used as a counter electrode. Calomel electrode was used as a reference electrode. The exposed area of the sample was 1cm^2. Potentiodynamic measurements were made at sweep rate of 5 mV s^{-1} from E_c – 0.15 V to E_c + 0.5 V. During the impedance measurements sinusoidal signal with amplitude of 10 mV was used. A spectrum was recorded at the open circuit potential in the frequency range from 0.01 Hz to 0.1 MHz.

The wettability of the obtained CCs was evaluated by sessile drop method on drop shape analyser DSA100 (KRÜSS, Germany). Solution of 0.5 M NaCl in bidistilled water was used for CA measurements. Wear of CC was tested by "ball-on-disk" method on tribometer Revetest-RST (CSM Instruments, Switzerland) using Al_2O_3 ball of 10 mm in diameter at a line speed of 50 mm s^{-1} and load of 10 N.

The surface morphology and cross-section of the CCs were studied with a scanning electron microscope (SEM) EVO 40 (Carl Zeiss, Germany).

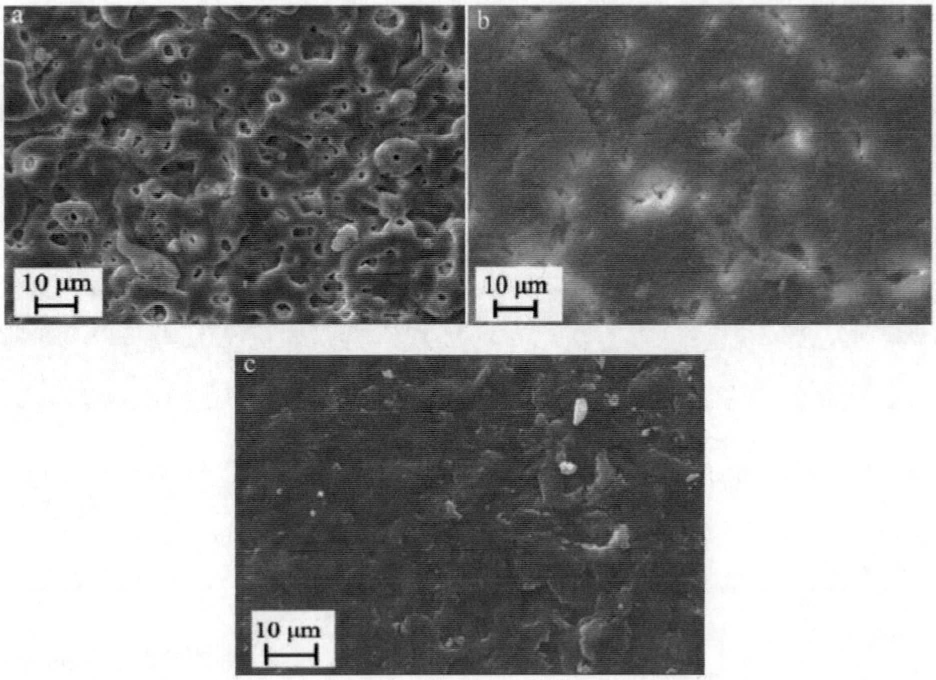

Fig. 1: SEM-images of **(a)** PEO-coating and composite coatings obtained **(b)** using telomeric solution and **(c)** using suspension of SPFE.

3.0. Results and Discussions

Data analysis of SEM images identified the influence of the multiplicity applying of organofluoric material and subsequent heat treatment on the surface morphology of the formed layers (Fig. 1). At surface treatment of the base PEO-coating (Fig. 1(a)) with the telomeric solution the number of pores and defects on the surface decreases and the surface becomes more uniform (Fig. 1(b)). In the case of applying SPTFE (Fig. 1(c)) a uniform film was formed without visible defects. According experimental date, the base PEO-layer consists of the different chemical compounds: MgO, $MgSiO_4$, MgF_2. From the analysis of cross-section images (Fig. 2) it can be concluded that particles of organofluorine materials as a single phase incorporate into the outer porous part of the PEO-layer, seal pores and microdefects in base PEO-coating. The thickness of CC (Fig. 2(b)) is nearest the thickness of base PEO-coating (Fig. 2(a)) due to obtain thin polymer-containing film on the PEO-layer.

The analysis of the data presented in Table 1 enable one to conclude that application of organofluorine material on the surface of the base PEO-coating has a positive effect on the protective properties of composite layers. In the case of applying the telomer of TFE on the surface the corrosion current density I_c reduces and the polarization resistance R_p increases by 4 orders of magnitude compared to the bare material and by 2 orders of magnitude compared to the PEO-coating (see Table 1). Formed coatings, which were obtained using of SPTFE suspension, have even higher protective properties. For such CC I_c reduces and R_p increases by 6 and 4 orders of magnitude compared to the uncoated alloy and PEO-coating, respectively (see Table 1). The conclusion can be done that CC-1 has lower values of R_p, and a higher I_c compared to CC-2 (see Table 1).

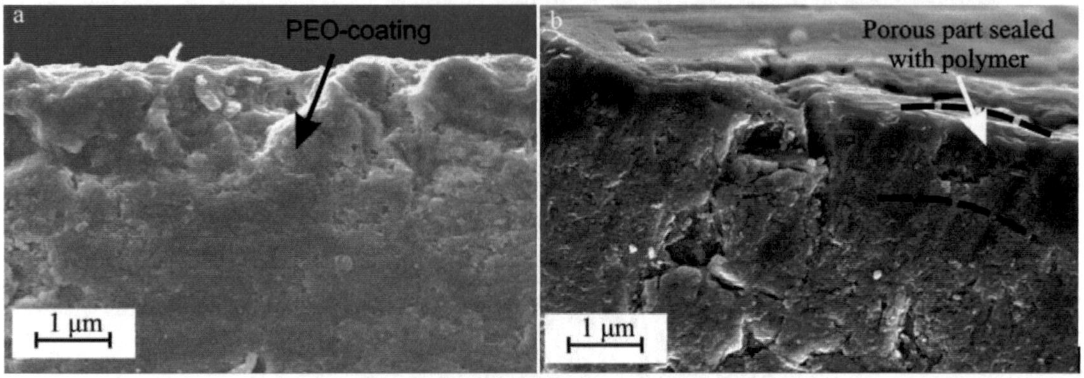

Fig. 2: Cross-sectional SEM-images of **(a)** PEO-coating and **(b)** composite coating.

Table 1: Corrosion and tribological properties and wettability of investigated samples

Type of coating	**E_c (V)**	**I_c (A cm^{-2})**	**R_p (Ω cm^2)**	**$\|Z\|_{f \to 0 Hz}$, (Ω cm^2)**	**CA (°)**	**Wear (mm^3 $(N\ m)^{-1}$)**
Mg alloy without coating	–1.61	5.3 10^{-5}	0.5 10^3	0.8 10^3	38±1	3.8 10^{-3}
Base PEO	–1.48	2.4 10^{-7}	1.0 10^5	4.3 10^4	49±2	1.7 10^{-3}
CC-1	–0.39	1.4 10^{-9}	4.6 10^7	5.1 10^7	168±2	5.7 10^{-6}
CC-2	–0.95	5.4 10^{-11}	4.9 10^8	1.9 10^9	140±2	7.6 10^{-7}

Apparently, this is due to the greater uniformity of the film obtained using a suspension SPTFE (Fig. 1(c)) compared to the film obtained using telomeric solution (Fig. 1(b)).

Data analysis of electrochemical impedance spectroscopy revealed a significant difference in the morphology of the CCs depending of the conditions of the heat treatment, the multiplicity of application and type of organofluorine material. Observed transformation of the position and amplitude of the time constants on the the plot of the phase angle θ versus the frequency f (Fig. 3b), which is indicating a change in the surface state: porosity, homogeneous of composite layers. In this paper, in order to describe the charge transfer mechanism on the coating/electrolyte interface, three kinds of equivalent electrical circuits (EEC) were used. For the uncoated sample the one-R-CPE-chain circuit was applied (Fig. 4(a)). For samples with coatings two kinds of EEC were used: two-R-CPE-chain circuit (base PEO-coating) (Fig. 4(b)) and three-R-CPE-chain circuit (composite coatings) (Fig. 4(c)). The using three-R-CPE-chain circuit associated with the appearance of the third time constant on the spectrum (Fig. 3(b)). This is a consequence of the sealing air by polymeric plug in the pores of PEO-layer. It should be noted that the dependence of the phase angle θ on the frequency f for CC-2 (Fig. 3(b)) is capacitive, which suggests a high uniformity of the obtained coating.

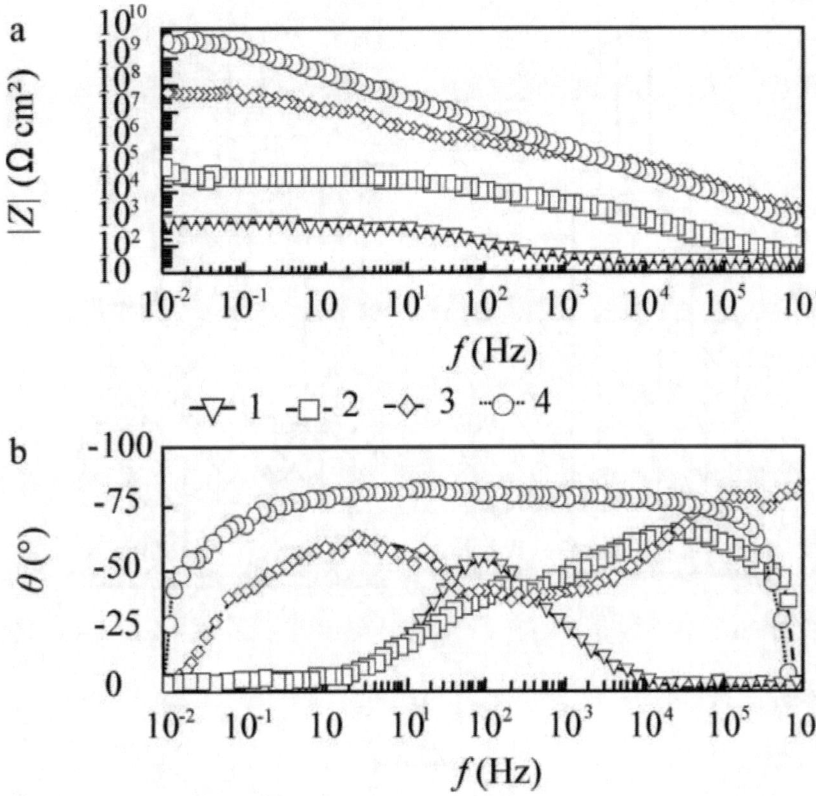

Fig. 3: Bode plots (dependence of the impedance modulus $|Z|_{f \to 0\ Hz}$ **(a)** and the phase angle θ **(b)** on the frequency f) for samples with a different surface treatment: 1 – without coating; 2 – PEO-coating; 3 – CC-1; 4 – CC-2.

The results of experiments enable one to conclude that multi-step treatment of the PEO-coated sample with telomeric solution increases the impedance modulus $|Z|_{f \to 0\ Hz}$ up to 10^7 Ω cm^2, and in the case of the treatment with suspension of SPTFE up to 10^9 Ω cm^2, that is almost by 3 and 5 orders of magnitude higher than one for the PEO-coating (Fig. 3(a), Table 1). Apparently, all above mentioned reasons of improving the characteristics of CC associated with the sealing the pores and defects, which are presented in the PEO-coating.

Results of wettability investigations for samples revealed the high hydrophobic properties of CCs (see Table 1). The highest values of the CA (168°) were obtained for the CC-1. Values of the CA greater than 150° indicate that this sample has superhydrophobic properties. In accordance with the results of previous studies[5] a superhydrophobicity is caused by multimodal roughness of surface. Apparently, in the case of using of the TFE telomeric solution such structure is formed as a result of the partial destruction of the organofluorine compounds and subsequent aggregation during heat treatment.

Analysis of tribological data revealed a positive impact of fluoropolymer particles implementation in the coating to improve the wear resistance of composite layers (see Table 1). Thus, for CC-1 and CC-2 wear reduces by 3 and 4 orders of magnitude compared to the base PEO-coating.

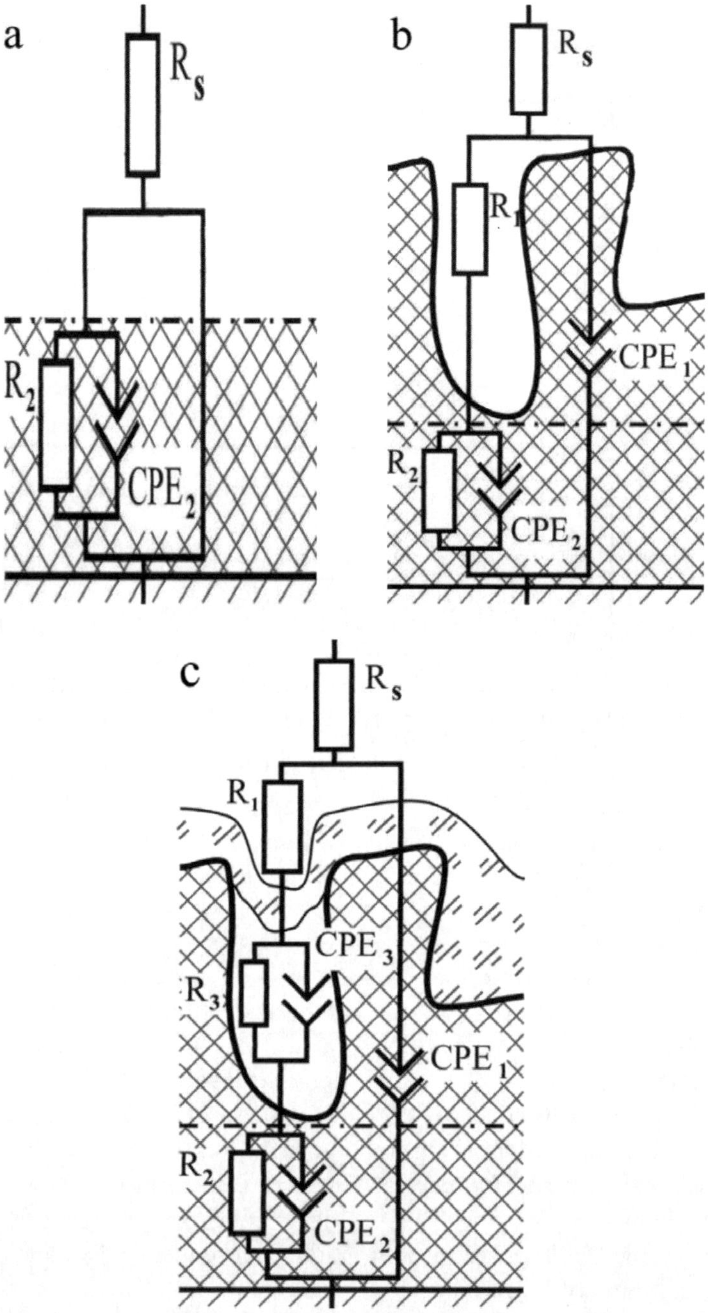

Fig. 4: Models of the composite coatings structure with the equivalent electrical circuits used for the fit of the experimental impedance spectra: **(a)** one-R-CPE-chain EEC; **(b)** two-R-CPE-chain EEC; **(c)** three-R-CPE-chain EEC.

Such increasing of the wear resistance is associated with the formation of the polymer-containing layer in the outer part of the PEO-coating, which has a low friction coefficient.

4.0. Conclusion

New method for the formation of the protective composite layers by PEO and subsequent treatment with fluoropolymer materials has been developed. The obtained polymer-containing coatings decrease the corrosion current densities and wear by 6 and 4 orders of magnitude, respectively, compared to unprotected alloy. Values of the contact angle attain to 168°.

5.0. Acknowledgments

This work was supported by the Russian Science Foundation (№ 14-33-00009) and the Government of Russian Federation (Federal Agency of Scientific Organizations).

6.0. Reference

1. Dunleavy, C.S.; Golosnoy, I.O.; Curran, J.A.; Clyne, T.W.; Surf. Coat. Technol., 2009, vol. 203(22), p. 3410.
2. Zhang, D.; Dong, G.; Chen, Yi.; Zeng, Q.; Appl. Surf. Sci., 2014, vol. 290, p. 466.
3. Kiryukhin, D.P.; Kim, I.P.; Bouznik V.M. Russian Federation Patent, 2010, 2381237.
4. Tsvetnikov A.K., Uminskiy A.A. Russian Federation Patent, 1992, 1775419.
5. Gnedenkov S.V., Egorkin V.S., Sinebryukhov S.L., Vyaliy I. E., Pashinin A.S., Emelyanenko A.M., Boinovich L.B. Surf. Coat. Technol., 2013, vol. 232, p. 240.

A Novel Surface Coating Technique for Ti Alloys

S. Miao,
Division of Materials,
School of Mechanical and Mining Engineering,
The University of Queensland,
St Lucia, Brisbane, QLD 4072, Australia.

Y. Shi,
Shenyang National Laboratory for Materials Science,
Institute of Metal Research,
Chinese Academy of Sci-ences, Shenyang, 110016, China.

M. Zhang*
Division of Materials,
School of Mechanical and Mining Engineering,
The University of Queensland,
St Lucia, Brisbane, QLD 4072, Australia.
*Email: mingxing.zhang@uq.edu.au, e5mzhang@uq.edu.au

Abstract

A novel packed powder diffusion coating (PPDC) technique has been developed to improve the high temperature oxidation resistance and wear resistance of Ti alloys. Optical microscopy observations show that a thick coating is formed on both pure Ti and Ti-6Al-4V substrates after PPDC treatment. The thickness varies from 100 μm to 1295 μm depending on treatment temperature and time. Applying this technique to Ti-10Cr-2Fe-3Al and Ti-6Cr-5Mo-5V-4Al alloys, uniform coatings with thickness up to 1000 μm have also been successfully obtained. Cyclic oxidation tests carried out in a static air within temperature range from 800 °C to 1000 °C for 24 cycles show high oxidation resistance of Ti alloys after PPDC treatment, which is comparable with some Ni-base superalloys. Friction and wear tests at a load of 10-50N for 30 min indicate that the average friction coefficient and volume loss of the coatings is much lower than that of the Ti-6Al-4V substrate, exhibiting a better wear resistance. Therefore, the PPDC treatment can not only effectively prevent Ti alloy substrates from oxidation in air up to 1000 °C, but can also significantly improve their room temperature wear resistance. It is predicted that, after the treatment there will be strong potential to partially replace the Ni-base or Co-base superalloys with Ti alloys for high temperature applications, leading to dramatic reduction of cost.

1.0. Introduction

The lightness, high specific strength and excellent corrosion resistance are the major advantages of Ti alloys, which at-tract an increasing interest in wider industrial applications[1,2]. Currently, about 80% of all the Ti products are used in the aerospace industries. However, these alloys suffer from poor oxidation resistance due to the high chemical activity when served in high temperature environment and low wear resistance when used as abrasively stressed parts. This not only shortens their service life but also limits their wider applications in areas associated with high temperature and heavy wear. The currently most popular Ti alloys, such as Ti-6Al-4V alloy, can only be used up to 300 °C. Even the so called heat resistant Ti alloys, Ti1100 and IMI 834 alloys, can only be used below 600 °C[3,4]. In order to overcome these limita-tions and to enable the application of Ti alloys to higher temperatures, both the wear resistance and oxidation resistance of the alloys must be improved.

Although development of new alloys may solve the problems associated with Ti alloys, but it significantly raises the production costs, and also possibly decreases the ductility and toughness. Therefore, surface treatment has been recognized as a major emerging technology that can efficiently achieve this goal without altering the properties of the substrate.

Although the currently available coating techniques, such as PVD, CVD, thermal spray and cold spray etc. could pro-duce an additional layer with excellent properties on the substrate of Ti alloys, the bonding strength is a major concern due to the difference in thermal expansion coefficient[5,6]. Surface alloying leads to formation of coatings that integrate with substrates by a high bonding force. Compared with other surface alloying techniques, such as electron beam and laser surface alloying etc.[7,8], diffusion coating is a simple, practical and low cost technique that does not involve any ex-pensive equipment. Although diffusion coating has been successfully used in ferrous alloys and Ni-base alloys for many years[9,10], challenges are remaining as this technique usually requires high temperature to promote the diffusion of atoms, which may alter the microstructure of substrate and thus deteriorate the overall properties. In the past years, Zhang and co-workers[11,12] have developed a low temperature surface alloying technique, which is termed as Packed Powder Diffu-sion Coating (PPDC), for metallic materials. Following the success in applying the technique to Mg and Cu alloys the present work aims to extend the PPDC process to Ti alloys. Then, the coating will be comprehensively characterized in terms of microstructures, chemical composition of coating and its properties, including cyclic oxidation and wear resistance.

2.0. Materials And Methods

2.1. Materials

As-received commercial Ti alloys plates, including pure Ti (Grade 2), Ti-6Al-4V, Ti-10V-2Fe-3Al and Ti-6Cr-5Mo-5V-4Al, were cut into 10 × 10 × 10 mm small blocks. All faces of the blocks were ground on silicon carbide (SiC) paper up to grade 600, followed by ultrasonic cleaning and air drying.

The masteralloy[13] used in the PPDC process is the optimized powder mixture of Al+20 wt.%Ni. The commercial Al powder has average particle size of 45 μm, the nickel powder has average particle size of 1~5 μm, Additional 1 wt.%NH_4Cl was used as activator and 20 wt.%Al_2O_3 powder with average particle size of 45 μm was used as filler. All powders and the activator are thoroughly mixed for 60 minutes in a mechanical powder-mixer before use.

Table 1: Chemical composition of as-received Ti alloys used in the present work (wt.%)

Ti alloys	Al	Cr	Mo	V	Fe	Si	Sn	Zr	C	O	H	N	Ti
Pure Ti (Grade 2)	-	-	-	-	0.35	-	-	-	0.010	0.105	0.001	0.005	Bal.
Ti-6Al-4V	5.5-6.75	-	-	3.5-4.5	0.40	-	-	-	0.08	0.20	0.015	0.05	Bal.
Ti-10V-2Fe-3Al	2.6-3.4	-	-	9.0-11.0	1.6-2.2	-	-	-	0.05	0.13	0.015	0.05	Bal.
Ti-6Cr-5Mo-5V-4Al	3.93	5.7	4.7	4.81	0.08	0.028	-	-	0.025	0.20	0.015	0.05	Bal.

2.2. Packed Powder Diffusion Coating (PPDC)

Optimized treatment temperature and times (650 °C for 3-15 hours) were used, followed by air cooling to room temperature.

2.3. Microstructure analysis

The thickness of coatings produced on Ti alloys was measured on optical micrographs. The microstructure and chemical compositions of these coatings were examined using scanning electron microscopy (SEM) together with energy-dispersive spectroscopy (EDS) on JEOL 6460LA. X-ray diffraction was used to characterize the phase structures of coatings.

2.4. Cyclic oxidation test

In order to evaluate the oxidation resistance of the PPDC coatings, thermal cyclic oxidation tests were conducted in static air at 800-1000 °C for 24 cycles. Each cycle consisted of one hour immersion in static air in the furnace, followed by a 30-minutes air cooling down to room temperature. The coated samples were oxidized together with pure Ti and with commercial Ti-6Al-4V (in wt.%) alloy for comparison. All these specimens were placed in a quartz tube with both end closed through a 20 mm hole on cylinder wall. An electronic balance was used to measure the weight change. The sample weights were measured to 0.0001 g. the oxidation kinetics was monitored by measuring the weight gain as a function of oxidation time and also by measuring the total scale thickness after oxidation.

2.5. Friction and wear tests

In order to further characterize the PPDC coating, dry sliding wear tests were performed on an Optimol SRVIII oscillating friction and wear tester at room temperature (25 °C) in air with a relative humidity of 40-50%. The machine was equipped with a ball-on-disc contact configuration, which use the WC-Co balls with a hardness of HV-1750 as the counter friction pair. Prior to testing, all specimens were cut into 3 mm thick plate and mechanically polished through a level of #1200 grit paper, followed by ultrasonic cleaning and air drying. Detailed testing parameters are as follows: 2 mm oscillating stroke with a frequency of 5 Hz, and a load ranging from 10 to 50 N were applied for 30 minutes, wear volume loss were determined by formula proposed by Qu et al[14]. For comparison, the tribological behaviours of Ti-6Al-4V samples were explored under the same conditions.

3.0. Results

3.1. Microstructure and thickness

The cross-sectional optical micrographs of PPDC coatings produced on pure Ti substrates in the powder mixture at 650 °C for 3-15 hours are shown in Fig. 1.

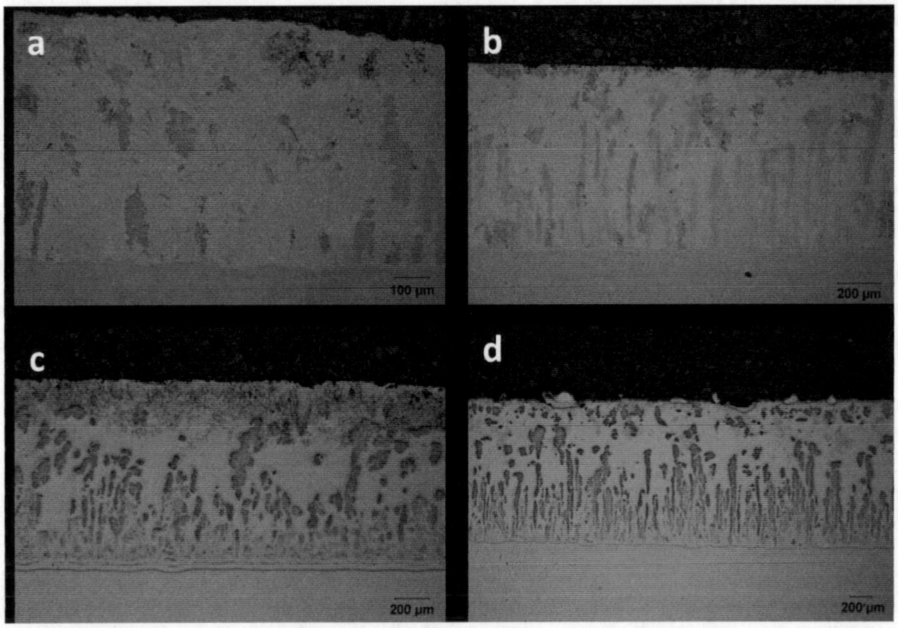

Fig. 1: Optical micrographs on the cross-sections of PPDC coatings on pure Ti substrate treated at 650 °C for different times: **(a)** 6 hours **(b)** 9 hours **(c)** 12 hours and **(d)** 15 hours

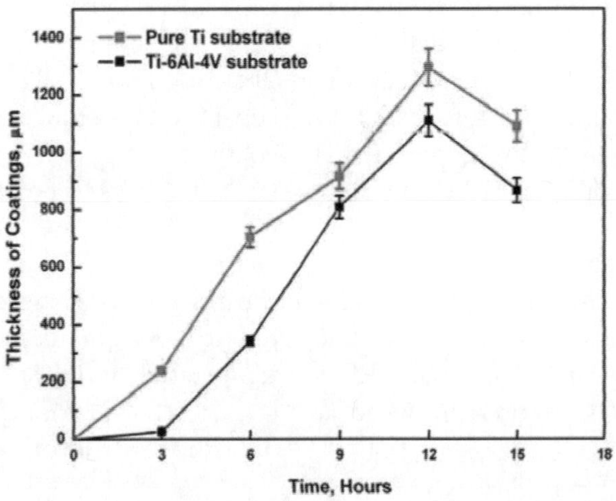

Fig. 2: Variation of the thickness of PPDC coatings on pure Ti and Ti 6Al 4V substrates with PPDC treatment time at 650 °C

All coatings produced for different times are integrated very well with the sub-strates and no cracks, gaps and pores are observed along the interface, indicating high bonding. One can also see that as the PPDC treatment time increases from 3 hours to 12 hours, not only the coating is thickened, but it also becomes more compacted. In the coating produced in 3 hours as shown in Fig. 1(a), tiny pores can be observed. However, fully compact coatings without noticeable pores or cracks are obtained after PPDC treatment is over 6 hours. Fig. 1(b)-(e) show dense coatings.

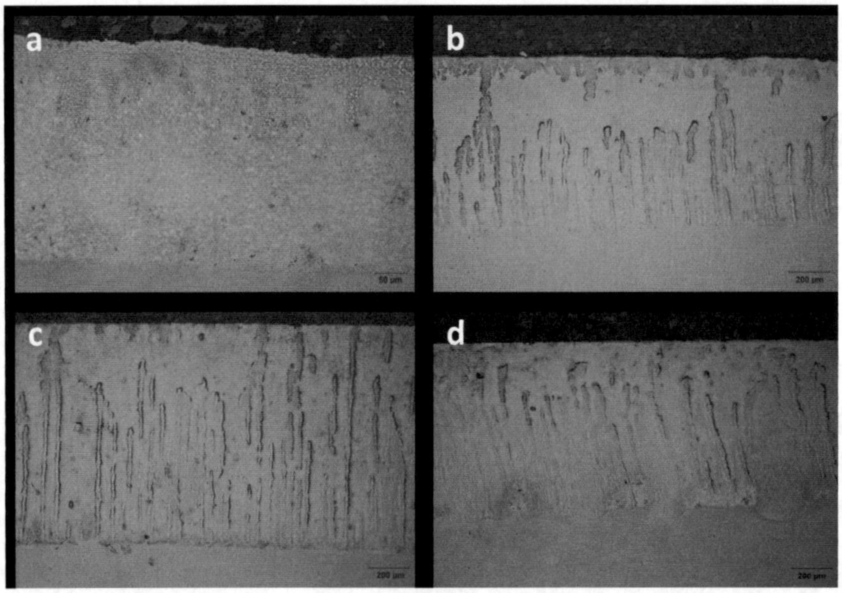

Fig. 3: Optical micrographs on the cross-sections of PPDC coatings on Ti-6Al-4V substrate treated at 650 °C for different time: **(a)** 6 hours **(b)** 9 hours **(c)** 12 hours and **(d)** 15 hours.

The variation of PPDC coating thickness on pure Ti substrates with the treatment time is plotted in Fig. 2. The thickest coating is obtained after 12 hours treatment. Treatment for longer time up to 16 hours reduces the thickness. The actual reason for this reduction in thickness is not clear. Some experimental errors may be responsible for the thinning. Further investigation is undertaking.

To valid the technique in Ti alloys, PPDC treatment was also carried out on Ti-6Al-4V alloy, which is a typical Ti alloy. The cross-sectional optical micrographs of PPDC coatings on Ti-6Al-4V substrates produced at 650 °C for 3-15 hours are shown in Fig. 3. Similar to pure Ti, all coatings exhibit high bonding with the substrates without cracks, pores and gaps. But, the dense coatings are obtained after PPDC treatment is over 6 hours. The variation of the coating thickness with the PPDC treatment time is also shown in Fig. 2. Compared to the coatings on pure Ti, the coatings produced on Ti-6Al-4V substrates are slightly thinner. It is considered that the existence of Al element in Ti-6Al-4V may suppress Al atom diffusion from the source powder into the substrate.

In order to further valid the techniques, PPDC coatings were produced on Ti-10V-2Fe-3Al and Ti-6Cr-5Mo-5V-4Al alloy substrates. As shown in Fig. 4, thick and dense coatings were also successfully obtained on both Ti alloys. Therefore, it is reasonable to conclude that the newly developed PPDC technique can be used to various Ti alloys and the thickness of the coating varies from 100 μm to 2000 μm depending on the treatment temperature and time.

3.2. Phase structure in the PPDC coating

To explore the phase structure in the PPDC coatings, back-scattered electron micrographs on the cross-sections of the coatings were examined in SEM. Fig. 5 shows a typical example of such micrographs, taken from the coating on pure Ti produced at 650°C for 15 hours. It can be seen that the coating consists of two different phases, forming composite coating.

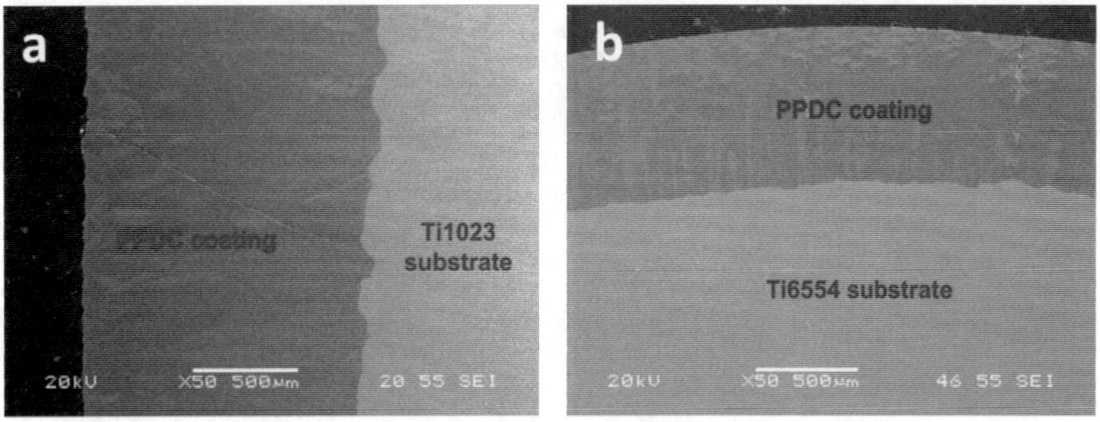

Fig. 4: SEM back scattered electron micrographs of the PPDC coatings produced on **(a)** Ti-10Cr-2Fe-3Al and **(b)** Ti-6554 substrates

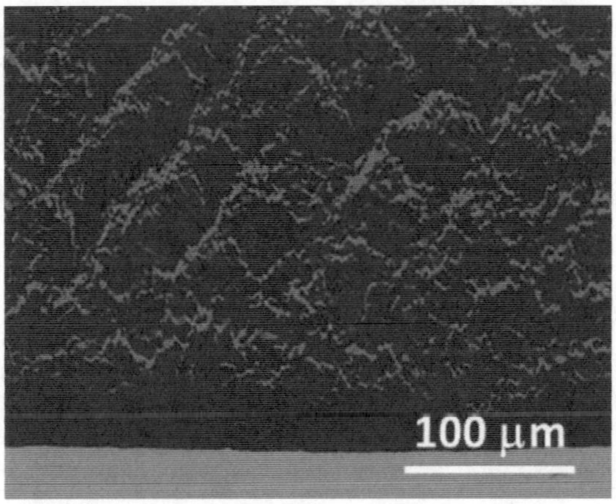

Fig. 5: SEM Back-scattered electron micrographs of coatings produced on pure Ti substrate produced at 650 °C for 15 hours

The "white" phase uniformly distributes in the "dark" matrix phase. Subsequent EDS analysis indicates that the "white" phase is Ni-Al intermetallic compound, and the "dark" phase is Ti-Al compound.

To further identify the phases in the coating, X-ray diffraction analysis was performed on both the topmost surface and on the coating layer close to the interface between the coating and substrate. The latter was explored through mechanical removal of materials outer layers from the substrate. The X-ray diffraction spectra are shown in Fig. 6. Indexing the spectra indicates that the coating contains TiAl3 and NiAl3 intermetallic compound and residual Al.

Compared to the monolayer of intermetallic, this composite coating could provide much more volume tolerance avoiding the formation of cracks during heating or cooling.

3.3. Oxidation resistance

Fig. 7 shows the kinetic curves of cyclic oxidation of PPDC coatings, pure Ti and Ti-6Al-4V in static air at 800 and 900°C for 24 cycles, respectively.

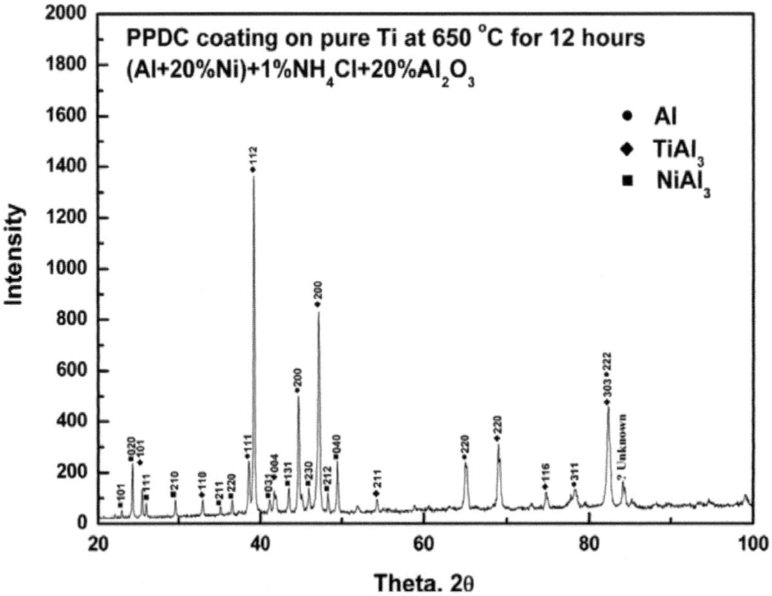

Fig. 6: X-ray diffraction patterns on the topmost surface and on the layer that is close the interface of PPDC coatings produced on pure Ti substrate at 650 °C for 15 hours

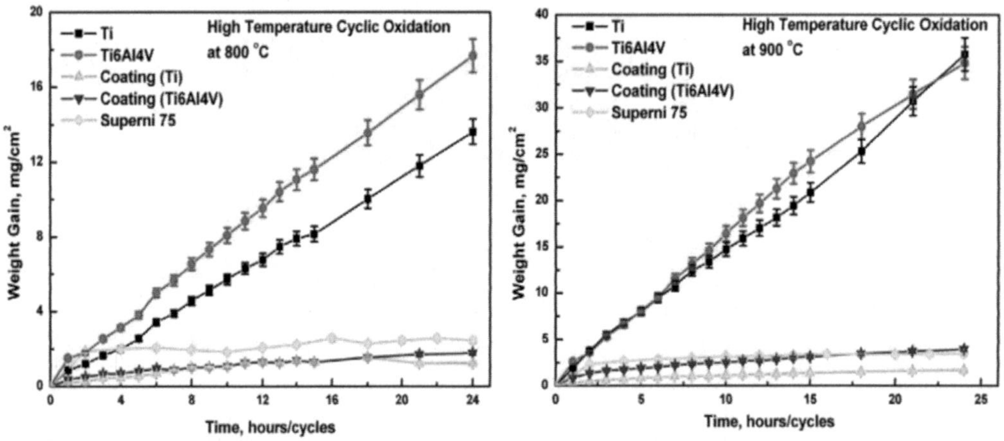

Fig. 7: Cyclic oxidation behaviours of the PPDC coatings on pure Ti and Ti6Al4V substrate compared with those of pure Ti, Ti6Al4V and Superni 75 alloys. Oxidation was undertaken at at 800°C and 900°C for 24 hours.

It can be seen that the weight gains of PPDC coating samples are much lower than that of pure Ti and Ti-6Al-4V at both temperatures. Compared to Ni-base superalloy Superni 75, the coatings show a comparable, even better oxidation resistance[15,16] than the Ni-base superalloy. This implies that some Ni-base superalloys could possibly be replaced with Ti alloys after PPDC treatment in industrial applications, therefore to significantly reduce the cost.

To investigate the effects of PPDC coatings with different thickness on their oxidation resistance, the weight gains of PPDC coatings produced at 650 °C for different time ranging from 3 to 15 hours have been measured at 1000 °C for 24 cycles and the results are shown in Fig. 8.

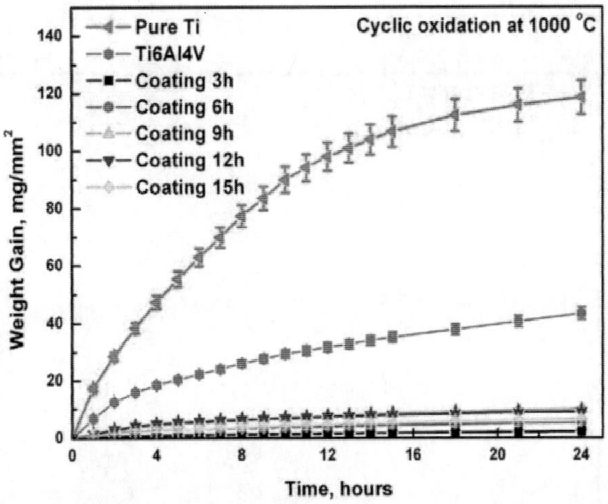

Fig. 8: Cyclic oxidation behaviors of the PPDC coatings on pure Ti and Ti6Al4V substrate compared with those of pure Ti and Ti6Al4V alloys. Oxidation was undertaken at 1000 °C for 24 hours

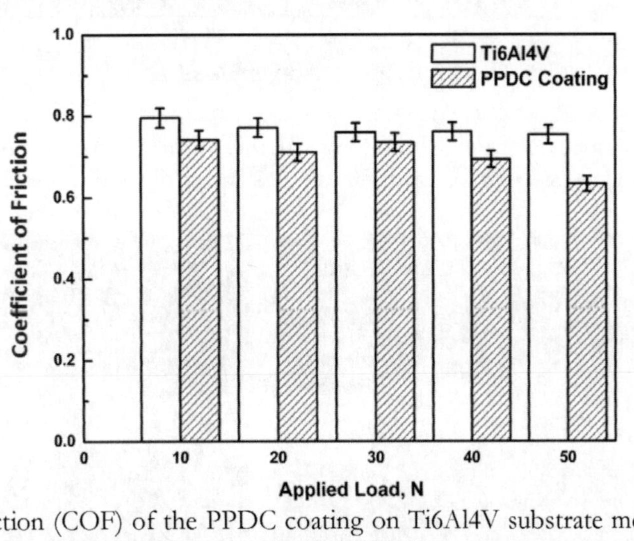

Fig. 9: Coefficient of friction (COF) of the PPDC coating on Ti6Al4V substrate measured at different loads for 30 minutes and compared to the Ti6Al4V alloy.

Compared to the pure Ti and Ti-6Al-4V, the PPDC coatings exhibit an excel-lent oxidation resistance, and the thickness of coating has no obvious effects on their oxidation behaviour. Due to the lack of available oxidation data for superalloys at 1000 °C, comparison of oxidation behaviour of the PPDC coating at 1000 °C with the superalloys cannot be made.

3.4. Friction and Wear

Fig. 9 shows the average friction coefficients of the PPDC coatings measured at different loads. At each load, the coat-ings shows lower friction coefficient than that of Ti-6Al-4V substrates. In addition, the PPDC coatings also exhibit higher wear resistance at dry sliding conditions, as shown in Fig. 10.

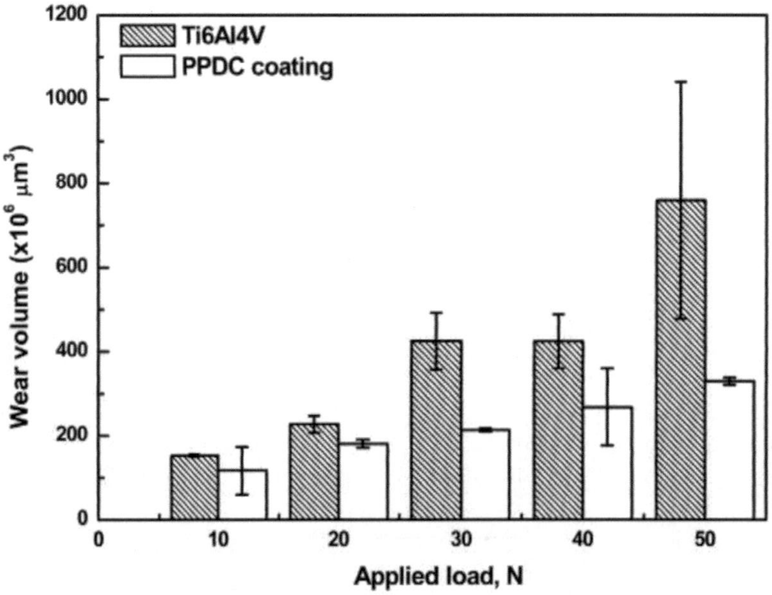

Fig. 10: Wear resistance, indicated by the wear volume, of the PPDC coating on Ti6Al4V substrate measured at different loads for 30 minutes and compared to the Ti6Al4V alloy

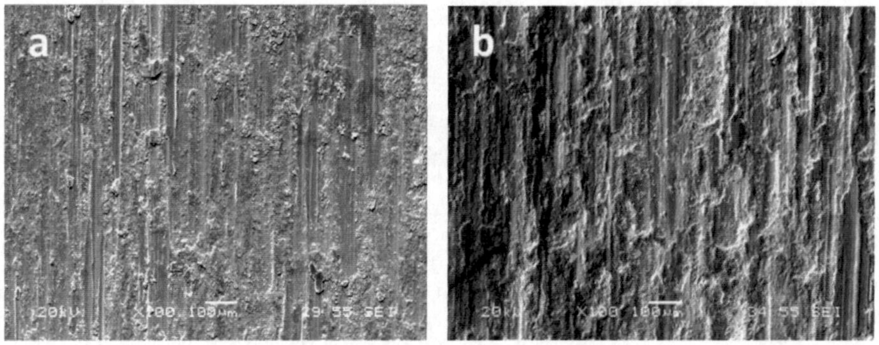

Fig. 11: SEM micrographs showing the morphology on worn surfaces after wear testing at a load of 50N: **(a)** Ti6Al4V **(b)** PPDC coating at 650°C

This is attributed to the hardening effect of $TiAl_3$ and $NiAl_3$ in the coating. After the dry sliding tests on both PPDC coating and Ti-6Al-4V substrate, although examination of the worn surface show the same abrasive wear, shallower groove can be observed on the worn surface of PPDC coating than that on surface of worn Ti-6Al-4V substrate, as shown in Fig. 11.

4.0. Conclusions

- Using the optimized PPDC process, thick coatings have successfully been produced on pure Ti and a few typical Ti alloys substrates, including Ti-6Al-4V, Ti-10V-2Fe-3Al and Ti-6Cr-5Mo-5V-4Al alloys.

- The thick coating produced on Ti alloy substrates is a composite coating, which consists of a compact $TiAl_3$ matrix with a distribution of $NiAl_3$ precipitates and a small amount of residual Al.
- The composite coating can significantly improve the oxidation resistance of both pure Ti and Ti-6Al-4V alloy at temperatures up to 1000°C. Compared to the Ni-base superalloy Superni 75 at both 800 and 900°C, the composite coating show a comparable or even better oxidation resistance than the Ni-base superalloys. This implies that some Ni-base superalloys could possibly be partially replaced by Ti alloys after PPDC treatment in industrial applications, and therefore to remarkably improve the fuel efficiency of aircrafts and thus reduce the cost.
- Compared to the as-received Ti-6Al-4V alloy, the composite coating also shows a higher wear resistance al-though both show the same abrasive wear behaviour. The enhanced wear resistance is attributed to the significantly improved hardness of composite coating.

5.0. Acknowledgments

The authors are grateful to Australian Research Council (ARC) for research funding and International Postgraduate Research Scholarship (IPRS) for scholarship support.

6.0. References

1. Boyer, R.; Welsch, G.; Collings, E.W.; Mater. Properties Handbook: Titanium Alloys, 1994.
2. Titanium and Titanium Alloys, WILEY-VCH Verlag GmbH & Co. KGaA, Weinheim, 2003.
3. Ti-1100: A new elevated temperature titanium alloy, SAMPE, Covinia, CA, 1968
4. Biallas, G.; Essert, M.; Maier, H.J.; Int. J. Fatigue, 2005, vol. 27, p. 1485.
5. Elsener, B.; Rota, A.; Bohni, H.; Mater. Sci. Forum, 1989, vol. 44-45, p. 29.
6. Nolan, D.; Huang, S.; Leskovsek, V.; Braun, S.; Surf. Coat. Technol., 2006, vol. 200, p. 5698.
7. Selvan, J.S.; Subramanian, K.; Nath, A.K.; Mater. Manuf. Process., 1999, vol. 14, p. 285.
8. Utu, D.; Marginean, G.; Pogan, C.; Brandl, W.; Serban, V.A.; Surf. Coat. Technol., 2007, vol. 201, p. 6387.
9. Harper, M.A.; Rapp, R.A.; Oxid. Metals, 1994, vol. 42, p. 303.
10. Jiang, S.M.; Xu, C.Z.; Li, H.Q.; Liu, S.C.; Gong, J.; Sun, C.; Corr. Sci., 2010, vol. 52, p. 435.
11. Zhang, M.; Reilly, K.; Kelly, P.; J. Mater. Res., 1999, vol. 14, p. 3271.
12. Zhang, M.-X.; Reilly, K.; Kelly, P.M.; J. Mater. Res., 2002, vol. 17, p. 2477.
13. Zhang, M.-X.; Miao, S.; Australian Provisional Patent Application 2011904192 "Method of Treatment" filed on 14 October 2011 by The University of Queensland.
14. Qu, J.; Truhan, J.J.; Wear, 2006, vol. 261, p. 848.
15. Sidhu, T.S.; Malik, A.; Prakash, S.; Agrawal, R.D.; Int. J. Phys. Sci., 2006, vol. 1, p. 27.
16. Singh, H.; Puri, D.; Prakash, S.; Maiti, R.; Mater. Sci. Eng. A., 2007, vol. 464, p. 110.

Influence of Selective Post-Deposition Laser Treatment on Tribological and Wear Properties of Cold Sprayed Titanium Coatings

A. Astarita,
Dept. Chemical, Materials and Industrial Production Engineering,
University of Naples Federico II,
P. Tecchio 80, 80125 Naples, Italy.

F. Rubino, P. Carlone*, A. Ruggiero,
Dept. Industrial Engineering,
University of Salerno, Via Giovanni Paolo II,
84084 Fisciano, Salerno, Italy.

C. Leone,
Dept. Chemical, Materials and Industrial Production Engineering,
University of Naples Federico II,
P. Tecchio 80, 80125 Naples, Italy.

M. Merola,
Dept. Industrial Engineering,
University of Salerno, Via Giovanni Paolo II,
84084 Fisciano, Salerno, Italy.

A. Squillace
Dept. Chemical, Materials and Industrial Production Engineering,
University of Naples Federico II,
P. Tecchio 80, 80125 Naples, Italy.
*Email: pcarlone@unisa.it

Abstract

The focus of this study is to assess the influence of a selective post-deposition laser treatment on the tribological and wear behavior of a titanium coating layer on aluminum. Titanium grade 2 powders were deposited onto an aluminum alloy AA2024-T3 sheet, by means of cold gas dynamic spray process. A selective laser post-treating was carried out by using a 220 W diode laser to enhanced wear properties of coatings. Tribo-tests were executed to analyze the tribological behavior of materials in contact with an alternative moving counterpart under a controlled normal load. Tribo-tests were performed on untreated aluminum sheets, titanium grade 2 sheet, as-sprayed titanium powders and laser-treated coating layer to compare the wear properties of different materials and the effectiveness of treatment. The results obtained proved the effectiveness of the laser treatment in improving the wear performances of the coating, in terms of both friction coefficient and mass lost. The laser treatment promotes a change of the wear mechanism, switching from a severe adhesive wear, resulting in the Galling phenomenon, to an abrasive wear.

1.0. Introduction

Titanium and titanium alloys are widely used in many industrial fields (e.g. aerospace, automotive, chemical industry, biomedical devices and so on) due to their high mechanical properties, excellent corrosion resistance and low specific weight. On the other hand, the massive use of titanium alloys is restricted by high costs of the raw materials and the manufacturing processes. The use of aluminum components, cheaper and lighter than titanium, coated with titanium powders, is an interesting solution. Aluminum alloys mainly used in aeronautics, e.g. the alloys of 2xxx and 7xxx series, are heat treatable alloys and age hardened. Thus, they require a deposition technique in which the substrate remains at low temperature during the whole process. The Cold Gas Dynamic Spray (CGDS) is an innovative deposition technology matching these requirements. CGDS is an additive manufacturing process used to create a coating layer by means of high velocity impacts of metallic particles dispersed in supersonic gas flows. In this process, coating deposition occurs at relatively low temperatures if compared to other spray technologies, allowing sprayed particles to preserve their solid state. In more details, appropriate flow conditions are enforced combining a high pressure heated gas with a converging/diverging nozzle (De Laval nozzle). Metallic particles are propelled to supersonic velocity (i.e. 500-1000 m/s), then the high kinetic energy causes the impingement of the particles onto the substrate[1]. The macroscopic plastic deformation, induced by the high velocity impact, was indicated by some researchers as the main bonding mechanism[2,3]. Two different scenarios have been observed, depending on materials pair and particles velocity: a) particles rebounding from the substrate or b) particles bonding to the substrate[4]. The velocity at which bonding is achieved is referred to as critical velocity and relies on the particle size and distribution as well as the substrate material. The coatings exhibit high density and conductivity, good corrosion resistance and high hardness, due to the cold worked microstructure[5]. Recent studies discussed the capabilities of post deposition laser treatments to compensate some process drawbacks, reducing the porosity, improving the superficial hardness and increasing anti-corrosion and tribological properties of the coating[6–8]. Several post deposition treatments were successfully used to modify the microstructure and to improve the properties of the cold sprayed titanium layer. Marrocco et al. (2011) proved that post-deposition laser treatment on titanium coating eliminates the residual micro-porosity and performes a high quality corrosion barrier layer[3]. Authors' previous work showed the optimal process parameters of laser treatment to obtain a compact titanium dioxide layer on cold-sprayed titanium coating[9].

The aim of this paper is to investigate the influence of laser treatment on the wear and tribological properties of titanium cold spray coating. Current developments in Ti processing predict the accessibility of lower-cost Ti and have revived interests in exploring the tribo-behavior of Ti alloys as bearing materials[10]. Titanium and its alloys exhibit low tribological properties. Their wear resistance can be improved by surface treatments promoting growth of the surface hardness, leading to changes of the wear mechanism and lowers the wear rate[11]. Frictional contact of titanium alloys both against other materials and titanium alloys themselves, especially under pure sliding conditions, quickly damages the contact surface area and results in the transfer of some particles of the material to the counter face[12,13]. Titanium and its alloys are characterized by a high and unstable friction factor, strong adhesion wearing, and high tendency to seizure (localized damage due to the diffusion welding between sliding surfaces[14]) and scuffing (damage characterized by surface unevennesses and asperities called hillocks[15]). Tribological characteristics of titanium alloys can be improved by different kinds of treatment. Bell and Dong (2000) proved that the oxidation treatment of Ti6Al4V performed a significant improvement in wear resistance, but there is no detailed analysis focusing on the application of the laser treatments on cold sprayed coatings. Friction and wear tests were performed on untreated aluminum sheets, titanium grade 2 sheet, as-sprayed titanium powders and laser-treated coating layer to compare the wear properties of different materials and investigate the effectiveness of treatment.

2.0. Materials and Methods

Commercially pure grade 2 titanium particles with an average size of 40 μm were deposited onto 3 mm thick AA 2024-T3 plates. The chemical composition and the main properties of both grade 2 titanium and AA 2024 are available in literature[16,17] and not here reported in the interest of brevity. A cold spray facility DYMET403J (Obninsk-Center for Powder Spraying), assembled with a round-shaped exit nozzle, with a final diameter of 4.8 mm, was used for spraying. Helium, employed as carrier gas, avoided the oxidation of titanium particles during the deposition process. The particles were sprayed at a velocity of 680 m/s with Helium as carrier gas, the gas temperature and pressure were kept at 600 °C and at 12 bars. A superficial layer of cold sprayed titanium powder 5 mm thick was obtained. The successive milling process allowed to obtain the surface finishing required for the following laser treatment and tribo-tests. After the milling process, a 2.5 mm thick titanium coating with a good superficial finishing was obtained. Laser treatments were carried out using a diode laser (IPG DLR-200-AC). The laser power is transferred via an optical fiber to the laser head and focused/defocused on the plate surface by means of a focusing lens (corresponding to a spot diameter of about 2 mm). Previous work[9] gave necessary information about optimal process parameters to obtain a fully dense and compact titanium oxidized layer on cold sprayed coating surface: scan speed and power of laser beam were 50 mm/min and 200W.

Friction and wear tests were carried out using a ball-on-flat testing apparatus on a Reciprocatory Friction Monitor (Ducom Instruments, Bangalore, India). A titanium ball was hold in contact against a flat specimen of the studied materials, aluminum alloy, titanium grade 2, titanium as-sprayed coating and laser treated layer. The ball was a 3 mm diameter Ti6Al4V, the specimens were cut with dimensions of 25x25x5 mm. The reciprocating movement was imposed to the sphere holder by a stepper motor, which swing a portion of the complete round back and forth. The spinning motion was therefore converted to a linear movement by a rocker arm. The friction force was constantly monitored using a load cell, and the evolution of the coefficient of friction during the test was recorded. A gravimetric analysis was executed to evaluate the wear, in term of mass loss. Before each test the specimen and the ball were cleaned, using ethanol and compressed air, then weighted on a precision scale ((accuracy of 0.01 mg). These tests were not performed in full compliance with the provisions of Test Method G133, Procedure A. The normal force in these tests was 12.0 N, instead of 25.0 N, and the time period was 20 min, instead of 16min and 40 s as prescribed by the standard. All other provisions of Test Method G133 were followed. The stroke length was 10 mm and frequency oscillation 5 Hz (harmonic alternative motion). Every test was realized at room temperature, in laboratory air at controlled levels of relative humidity. A topographic analysis was carried on a 3D non-contact optical profiler, PLu S neox. The worn surfaces, cleaned from debris, were scanned using a magnification of 5x.

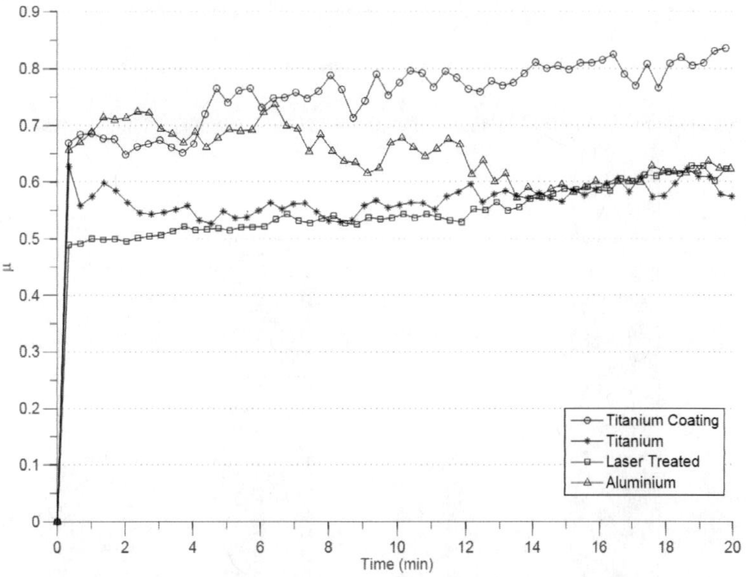

Fig. 1: Friction coefficients evolution

The scans yielded isometric and contour images, which provided rapidly qualitative information on the worn area and also some quantitative information, such as the maximum wear depth.

Microscopical analysis were also performed, employing Scanning Electron Microscopy (Hitachi TM 3000), to observe macroscopic features (depth and width) of the laser treated zone as well as to investigate microstructural aspects. Kroll's reagent (2 ml HF, 6 ml HNO3 and H2O up to 100 ml) was adopted to unveil microstructure, grain boundaries and phase distribution. Finally, Vickers microhardness tests were executed onto laser treated specimen to characterize the mechanical properties of material.

3.0. Results and Discussion

In Fig. 1 the kinetic coefficient of friction (μ) evolution for the different tribopairs is represented. Only one representative curve for each tribopair was reported in the interest of image readability.

The aluminium plate, coupled with the titanium ball (hereinafter omitted for simplicity), presents an almost steady state of the μ coefficient, with a value of approximately 0.6, after 14 minutes of test. The oscillation of μ in titanium plate is less evident, and all along the test duration it was included in the range of 0.5 – 0.6. Different discussion should be made for the titanium coating, in which the μ exhibits the less stable evolution. Even after 14 minutes the variations are still noticeable (in the range of 0.75 – 0.85). Finally, the evolution of the laser treated coating presents an increasing trend until reaching its maximum and almost stable position in the final phase, corresponding to 0.62. In the same phase of the test, apart from the titanium coating, the μ does not show large variances among the different materials.

The wear weight, evaluated by gravimetric analysis, yielded the results summarized in Fig. 2, in which both the values of the specimen and the sphere wear are shown and summed up. Regarding the aluminium, the wear of the ball was almost null, whereas the specimen lost a significant quantity of mass. Laser treated coating, yielded the highest ball wear of the different tribopairs.

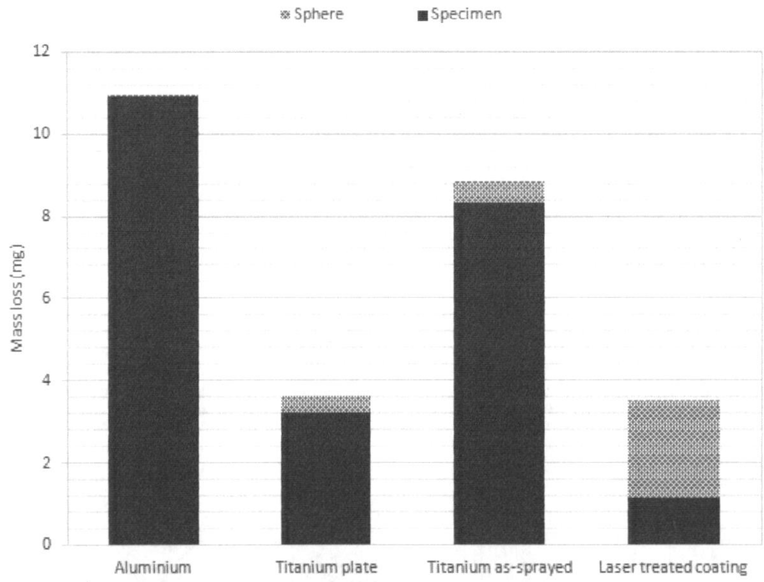

Fig. 2: Wear comparison in term of mass loss

This can be well-explained considering the high hardness of the titanium surface after the laser operation, which is responsible for a severe abrasive wear on the softer ball material. The wear mechanism itself changed due to the laser treating, switching from adhesive to abrasive wear. On the other hand, this specimen lost a tiny quantity of mass. The overall wear of the laser treated coating is comparable to the one of the titanium plate, in the latter case though, the main loss was observed on the specimen side.

In Fig. 3 is represented one loop cycle of the frictional force for each specimen type. The force was recorded by the load cell with a frequency equal to 100 times the oscillating frequency of the sliding ball. The curves represent a single evolution of the frictional force during the rubbing of the counterbody on the specimen surface, moving from – 5 mm to 5 mm, recorded in the final phase of the test. In Fig. 3(a) the Aluminium case is shown, the cycle is quite irregular; this unexpected behavior can be, in first analysis, ascribed to an uneven density of the specimen in closeness of the dead end center of the stroke. In Fig. 3(b) the force evolution cycle relative to the titanium plate is shown. In this case, the force has an irregular shape, but exhibits the typical hysteresis appearance proper of this evolution. Rather, having two peaks in correspondence of the two dead centers, where the ball diverts its movement and therefore the friction switches from the kinetic to the static phase. The loop cycles of the last two cases, i.e. the titanium coating (Fig. 3(c)) and the laser treated coating (Fig. 3(d)), are defined by the hysteresis evolution. These cycles, unlike the first two, exhibited a regular and flat shape during the sliding of the ball in between the two dead points. As already mentioned in the analysis of the friction coefficient (in Fig. 2), the mean value of the friction force of the titanium coating is higher than in the other cases.

In Fig. 4 the images obtained from the topography are displayed. The aluminium wear track (Fig. 4(a)) presents an irregular shape: it is non-symmetric in respect to the X-axes (orthogonal to the sliding direction).

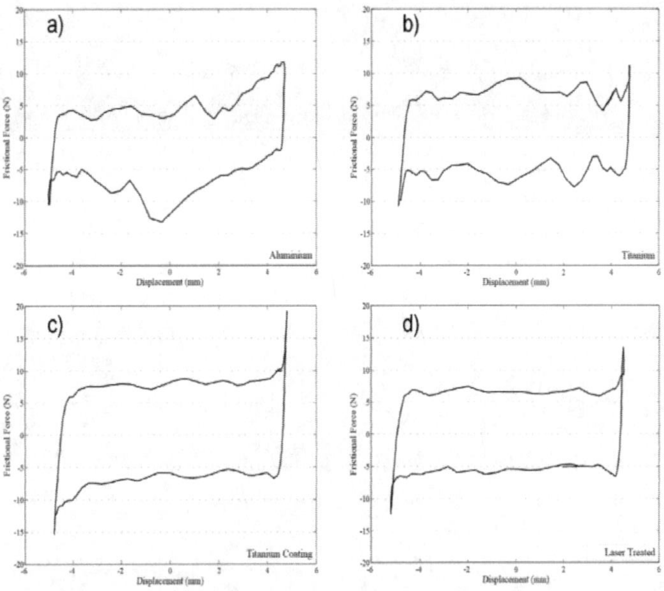

Fig. 3: Loop cycles of the frictional force **(a)** aluminium plate **(b)** titanium plate **(c)** titanium as-sprayed **(d)** laser treated coating

This observation is congruent with the analysis made of the friction loop cycle, which also showed an irregular profile. The worn track is deeper in correspondence to the end dead center (the right side of the track), where, in fact, the friction force was higher. The Titanium plate exhibits also an irregular shape (Fig. 4(b)), with the deepest points of wear in proximity to the two dead centers of the ball. It is also noticeable an area of strong plastic deformation along the border of the wear track (the red points of the contour image). In Fig. 4(c) the worn track of the Titanium coating is showed, the profile is regular and smooth, presenting the deepest wear of the specimens. Fig. 4(d) shows the topography result on the laser treated coating worn surface, as expected the image does not allow to have a proper estimation of the wear depth, due to the irregular profile yielded during the laser operation. As consequence of this treatment, the surface shows an indented profile and the worn area is distinguishable only thanks to the presence of a large boundary incision.

Micrographies of laser treated specimen (Fig. 5) show a dense titanium oxide layer produced on the surface of the titanium coating. In the center of the treated sample the oxide penetrates the coating up to about 30% of its thickness. Below the oxide layer, the heat produced by laser source caused the formation a heat affected zone, characterized by a microstructure with coarser grain than the material base.

Fig. 6 shows the different microstructures produced by the laser treating. Three different zones are visible in the sections of the treated specimen: the oxided zone (Fig. 6(a)), in which the typical microstructure of the titanium oxide is appreciable, the heat affected zone (Fig. 6(b)), which is characterized by lamellae of coarser dimensions than the as-sprayed titanium due to the heat input occurring during the laser process, and the base material (Fig. 6(c)), characterized by lamellae of coarser dimensions.

EDS analyses confirmed the existence of a rutile (TiO_2 titanium dioxide) layer.

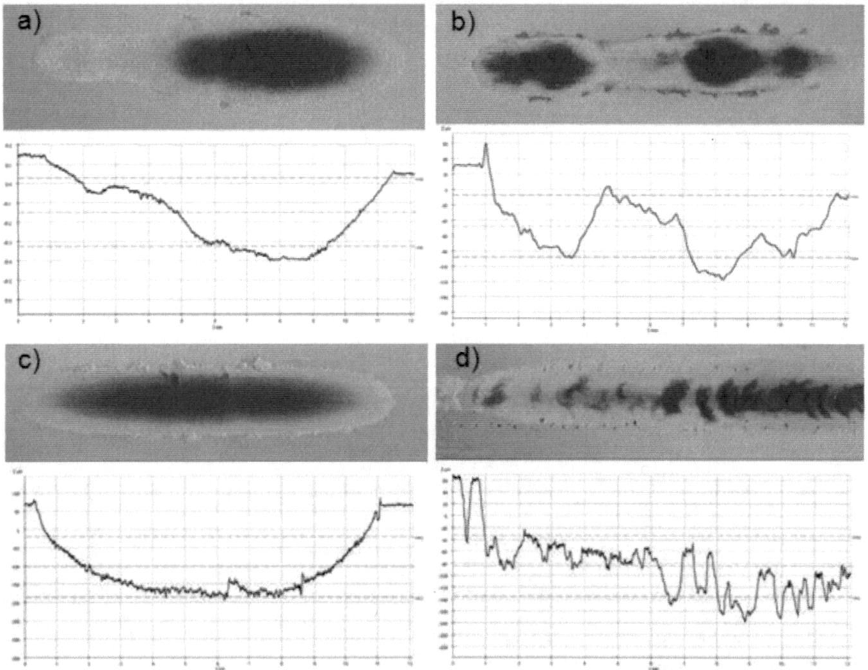

Fig. 4: Contour and profile in the midpoint of the wear tracks **(a)** aluminium plate **(b)** titanium plate **(c)** titanium as-sprayed **(d)** laser treated coating

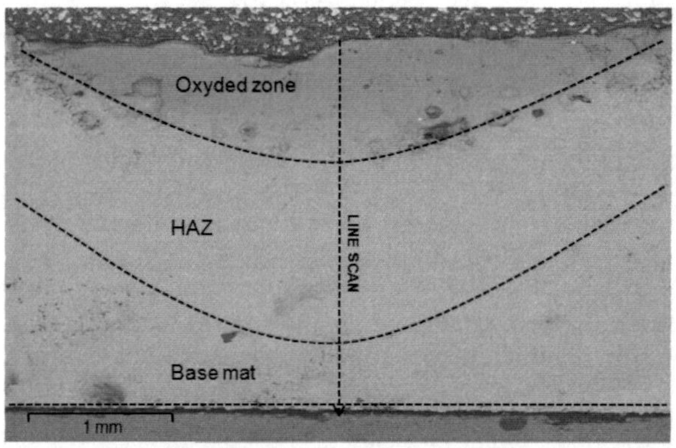

Fig. 5: Cross section of laser treated specimen (magnification 60x)

Line scan EDS analysis (Fig. 7), performed in the center of cross section following the pattern labeled "line scan" represented in Fig. 5, revealed the concentration of titanium and oxygen from the top of the surface to interface between coating and substrate.

The relative concentration of two elements, in the top of cross section, is approximately equal to stoichiometric one for the titanium dioxide; moving to the interface with the aluminum substrate, the oxygen percentage decreases to characteristic values of titanium grade 2.

The micro-hardness values of treated materials on the coating surface are about 1200 HV, much greater than the values of cold-sprayed titanium(about 200 HV) reported in literature[18] (Fig. 8).

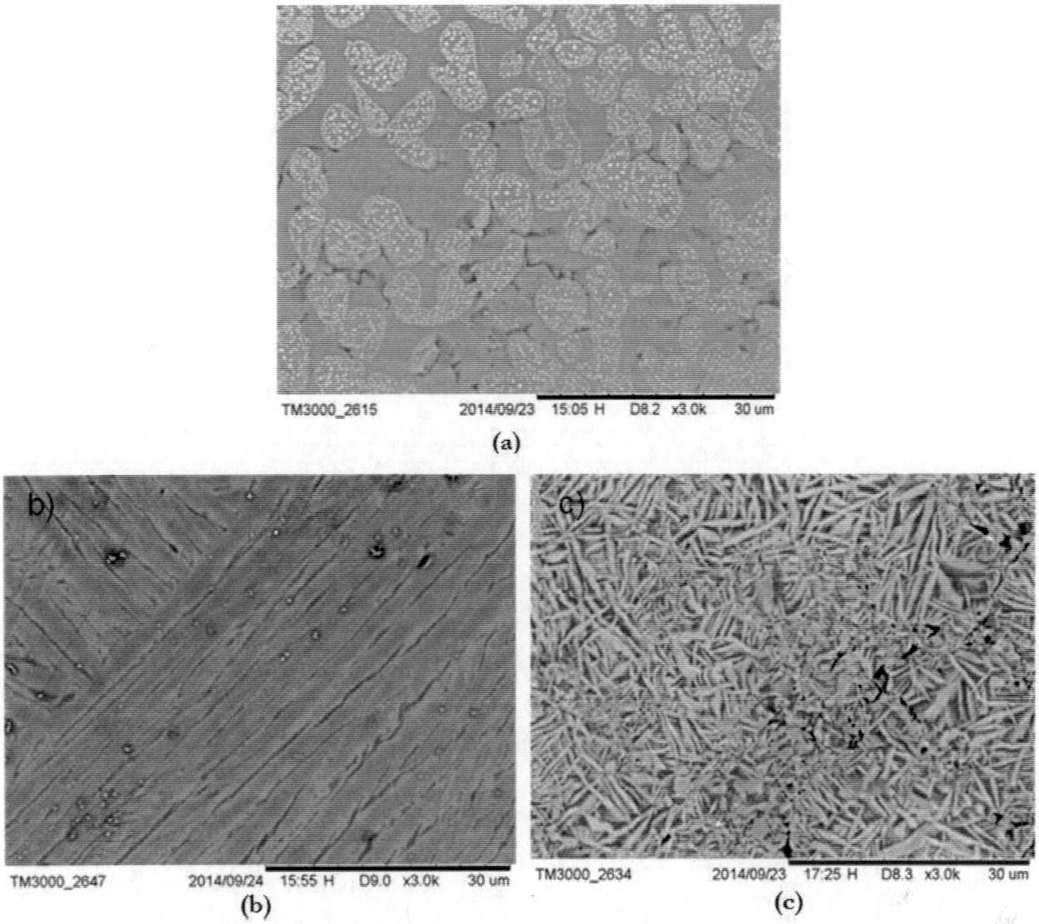

Fig. 6: **(a)** titanium oxided zone **(b)** heat affected zone **(c)** base material

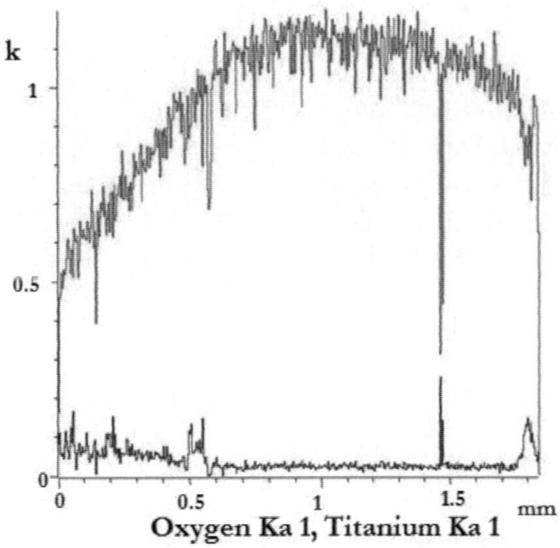

Fig. 7: Titanium and oxygen intensity along the scan line indicated in Fig. 4

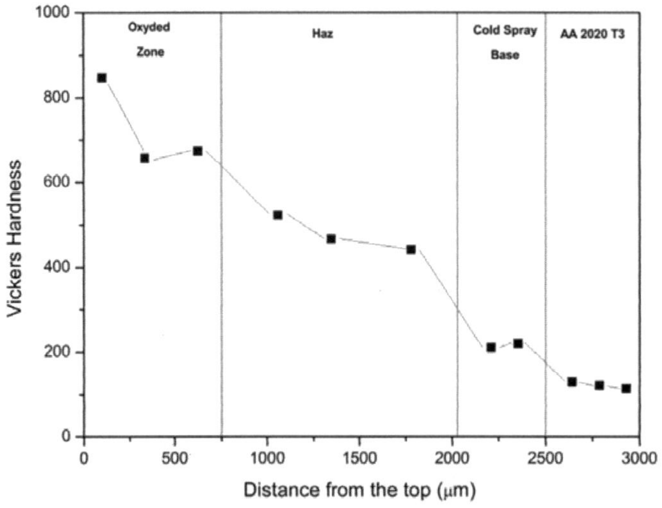

Fig. 8: Microhardness measurements

The tests were performed following same pattern of the EDS analysis.

Microhardness measurements proved the formation of a very hard layer. This observation is in agreement with the results of the tribological test (Fig. 2), which showed the major mass loss of the titanium ball than the laser treated titanium.

4.0. Conclusions

The following conclusions can be highlighted:

- Laser processing is an effective treatment to produce a harder oxidized layer on the top surface of the titanium coating.
- EDX analysis confirmed the oxidization of the titanium, evidencing also an increase in the Oxygen content within the treated area.
- Three different metallurgical zones are visible in the section of the treated tracks: base material, heat affected zone and the treated zone
- The temper state of the aluminium substrate was not affected during the process.
- The hardness of the titanium coating was enhanced by the laser treatment, achieving a superficial hardness of approximately 800 HV.
- The laser treatment influenced both the friction coefficient and the wear mechanism of the coating. The untreated coating showed an adhesive wear with a noticeable mass loss from the coating itself and a negligible wear of the ball. Abrasive wear mechanisms were observed testing the treated coating, with a negligible wear of the coating but a severe wear and mass lost on the sphere.
- Laser treatment improved the wear performances of the titanium cold sprayed coating.

5.0. References

1. Schmidt, T.; Gaertner, F.; Kreye, H.; J. Therm. Spray Technol., 2006, vol. 15(4), p. 488.
2. Morgan, R.; Fox, P.; Pattison, J.; Sutcliffe, C.; O'Neill, W.; Mater. Lett., 2004, vol. 58(7–8), p. 1317.
3. Marrocco, T.; McCartney, D.; Shipway, P.; Sturgeon, A.; J. Therm. Spray Technol., 2006, vol. 15(2), p. 263.

4. Astarita, A.; Durante, M.; Langella, A.; Montuori, M.; Squillace, A.; Surf. Int. Anal., 2013, vol. 45(10), p. 1530.
5. Irissou, E.; Legoux, J.G.; Arsenault, B.; Moreau, C.; J. Therm. Spray Technol., 2007, vol. 16(5-6), p. 661.
6. Bell, T.; Dong, H.; Wear, 2000, vol. 238(2), p. 131.
7. Wang, Y.; Li, C.; Guo, L.; Tian, W.; Surf. Coat. Technol., 2010, vol. 204, p. 3559.
8. Singh Sidhu, B.; Laser Surf. Modif. Alloys For Corr. Erosion Res., 2012, p. 355.
9. Rubino, F.; Astarita, A.; Carlone, P.; Genna, S.; Leone, C.; Memola Capece Minutolo, F.; Squillace, A.; Mater. Manuf. Proc., in press.
10. Qu, J.; Blau, P.J.; Watkins, T.R.: Cavin, O.B.; Kulkarni, N.S.; Wear 258 (2005), 1348-1356.
11. ASM Specialty Handbook: Aluminum and Aluminum Alloys.
12. Bloyce, A.; Morton, P.H.; Bell, T.; ASM Handbook, ASM Int. Materials Park, 1994, vol. 5, p. 835.
13. Bell, T.; Dong, H.; Proc. First Asian Conf. on Tribol., Ed. State Key Laboratory of Tribology, 1998, p. 421.
14. Dong, H.; Bell, T.; Ind. Lubrif. Tribol., 1998 ,vol. 50, p. 282.
15. ASM Handbook, Friction, Lubrication, and Wear Technology, vol. 18, p. 1879.
16. Astarita, A.; Durante, M.; Langella, A.; Squillace, A.; Metal Science and Heat Treatment, 2013, vol. 54 (11-12), p. 662.
17. Hanson, B.; The Selection and Use of Titanium: A Design Guide. Institute of Materials: Book Materials information service, 199), p. 641. ISBN 0901716979, 9780901716972
18. Wong, W.; Rezaeian, A.; Irissou, E.; Legoux, J.-G.; Yue, S.; Adv. Mater. Res., 2010, vol. 89-91, p. 639.

Thermochemical Treatments for Combating Extreme Wear – Boriding and Vanadizing

K. V. Dahl, A. E. Barington, T. L. Christiansen*
Technical University of Denmark,
Dept. Mechanical Engineering, Kgs. Lyngby, Denmark.
*Email:tch@mek.dtu.dk

Abstract

The present paper addresses the thermochemical treatments boriding, vanadizing and nitriding with subsequent vanadizing (nitrovanadizing) for improving the wear performance of Sverker 21 (AISI D2) tool steel. Vanadizing, nitrovanadizing and boriding of the Sverker 21 resulted in surface layer thicknesses of 2-4 μm, 4-6 μm and 30-40 μm, respectively. Nitriding prior to vanadizing resulted in formation of a V(C,N) carbo-nitride. Micro wear testing was used to simulate three-body abrasive wear with quartz as abrasive particles; all the treated samples had a higher wear rate than a cemented tungsten carbide reference. Vanadized samples performed the worst due to piercing of the thin layer by single wear particles. Nitrovanadized samples performed comparably to borided samples. Nitriding as a pre-treatment to vanadizing enhances the layer growth kinetics of the V(C,N) layer and is considered a promising way to improve wear resistance of vanadized components.

Surface Modification Technologies XXIX
Edited by T. S. Sudarshan and M. A. J. Somers

1.0. Introduction

Severe wear is a problem in many industrial applications. The material of choice for combating extreme wear is cemented tungsten carbide, which consists of hard carbides embedded in a soft metallic binder phase. However, there is a strong interest in materials solutions that can offer increased life times of wear parts whilst being cheaper than, for instance, hard-facings or claddings based on cemented carbides.

Boriding is a process where boron is diffused into the material resulting in a hard layer of borides of the elements present in the treated substrate. When the process is employed on steels iron borides are formed with a hardness of more than 1600 HV[1–3]. The borides formed are typically FeB and Fe_2B as well as borides of the alloying elements, e.g. Cr and Mo[1,2]. For tribological applications a single layer of Fe_2B is usually preferred since it is less brittle than FeB [2,4]. Boriding is a commonly used process for increasing wear resistance because of the high hardness combined with a low friction coefficient[1].

In vanadizing a hard carbide or nitride layer is formed by thermo reactive diffusion (TRD). The TRD process, also known as the Toyota process, was originally developed by Arai[5]. The substrate to be treated is typically immersed in a molten salt bath containing elements that readily form carbides or nitrides, in this case vanadium. The vanadium reacts with the carbon or nitrogen in the base material to form a thin layer of vanadium carbide or nitride. Vanadium carbide produced in this manner has a hardness of around 3500HV and has excellent adhesion to the substrate due to the nature of the process. Typical thicknesses are in the 2-14 μm range. Due to the low thicknesses the life time will be limited under conditions where significant wear of the hard layer occurs[6].

Casteletti et al.[7] tested both vanadized and borided D2 and H13 steel samples with a micro-wear tester with a fixed ball configuration without the use of an abrasive. They found vanadized samples to have superior wear resistance compared to borided samples. Compared to the untreated substrates both boriding and vanadizing resulted in large improvements in wear resistance.

Applying a nitriding, nitrocarburizing or carburizing treatment to the material prior to the TRD process results in a thicker carbide/nitride layer, without depleting the substrate in carbon. When nitrogen is present (nitriding and nitrocarburizing) a slightly lower hardness is obtained. Vanadium then reacts with both nitrogen and carbon resulting in a compositionally graded V(N,C) layer, which is carbon rich at the surface and nitrogen rich at the interface to the substrate[6,8,9].

In the present work boriding and vanadizing surface treatments with and without prior nitriding were applied on tool steel and evaluated against sliding abrasive wear from quartz particles and compared with the performance of a WC cemented carbide.

2.0. Experimental

The Uddeholm tool steel Sverker 21 (AISI D2) was chosen as substrate for the different surface treatments. The nominal composition of the Sverker 21 steel is shown in Table 1. A cemented tungsten carbide (WC) with a Co binder phase was used as a reference material.

Samples with a size of 10x10x8 mm were cut from the tool steel and 10x10x4 mm samples from the cemented tungsten carbide reference material. Before surface treatments, the steel samples were ground with SiC paper to 1000 grit before nitriding and 4000 grit before vanadizing or boriding.

Sverker samples were nitrided in the as-delivered soft annealed condition. Nitriding was done in a ceramic tube furnace, using ammonia as nitrogen source.

Table 1: Nominal composition for Sverker 21 tool steel in weight percent

C	Si	Mn	Cr	Mo	V
1.55	0.3	0.4	11.8	0.8	0.8

Before heating, the furnace was flushed with ammonia for 1 hour. Nitriding was then performed at 525°C with a linear gas flow of 0.255 m/min. NH_3. Heating to the nitriding temperature was performed in less than 60 minutes. The nitriding time was 550 minutes.

Boriding was done at Härterei Gerster AG and vanadizing was done at Chromin Maastricht BV; both treatments are commercially available processes. The samples were subsequently heat treated at 1030°C followed by quenching and double tempering at 500°C for two hours.

Half of the vanadized samples had prior to vanadizing been nitrided according to the procedure described above. After nitriding and before vanadizing, these samples were polished to 4000 grit for removal of the nitride compound layer. The hardness increase induced by the nitriding procedure is removed as part of the high temperature vanadizing treatment but the nitrogen is then available as reservoir for growing the hard outer V(C,N) layer.

Abrasive wear testing was performed on a micro-wear tester. The machine simulates 3 body isotropic abrasive wear by rotating a table and a sample holder while pressing them together with abrasive slurry in between. It is similar in function and design to many commercial polishing machines. The set-up is described in detail in Ref.[10]. Speed and pressure parameters were held constant throughout all the tests at 1 m/s and 100 kPa. Test duration was 20 seconds, resulting in a wear distance of 20 meters. Testing was done in a slurry composed of quartz mineral particles (size range 0.09-0.2 mm) suspended in 1,4-butanediol. The samples were then ranked according to an abrasion index, $\mathcal{A}_i$[10]:

$$\mathcal{A}_i = 10^4 M_N \,/\, (\mathcal{A} N \upsilon (2.38 + \varrho)) \quad [\mu m/m]$$

Where, $\mathcal{A}_i$ is the abrasion index, M_N is the weight loss after N cycles, $\mathcal{A}$ is the area of the sample, N is the normal pressure, υ is the speed and ϱ is the density for the worn part. The equation converts the measured weight loss to a height loss per travelled distance.

Before wear testing, cross sectional samples were cut from the surface treated steel samples and the cemented tungsten carbide. Standard metallographic methods were applied with a final polishing step of 1μm diamonds; Kallings reagent no. 1 (9.5g $CuCl_2 \cdot 2H_2O$ in 165 mL H_2O, 165 mL Ethanol and 165 mL HCl) was used for etching. Reflected light microscopy (RLM) of cross sections and worn surfaces was done using an Olympus CK40M light optical microscope equipped with a Leica DFC450C camera. Before SEM investigations the samples were sputter coated with a thin carbon layer using a Quorum Q150R Rotary-Pumped Sputter Coater. A JEOL JSM-5900, equipped with LaB6 filament and Oxford Instruments energy dispersive spectroscopy (EDS) was used for SEM investigations. For XRD analysis a Bruker D8 Discover diffractometer was used; step size of 0.04° and step time of 2s applying a conventional symmetric Bragg-Brentano geometry and Cr-Kα radiation.

3.0. Results and Discussion

3.1 Characterisation of materials prior to wear testing

The applied surface treatments resulted in the formation of hard layers on all samples. Fig. 1 shows cross sections of borided, nitrided, vanadized and nitrovanadized samples.

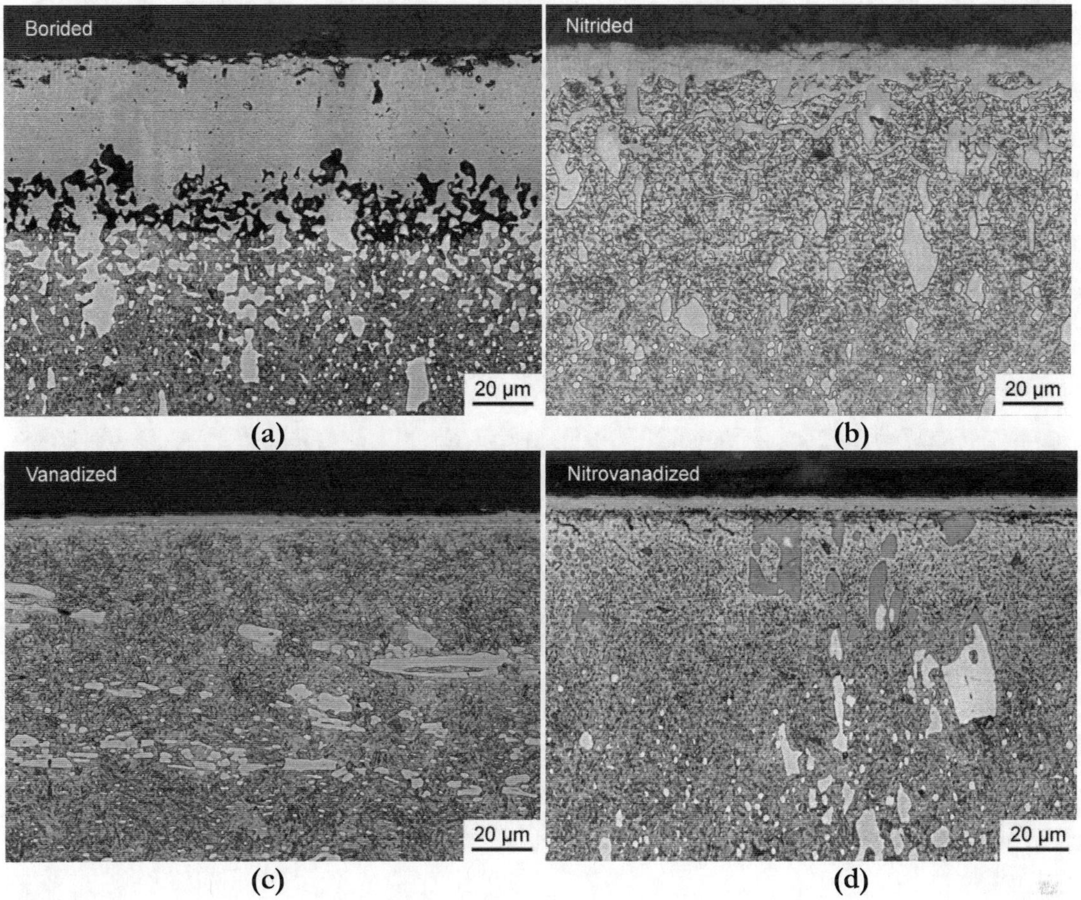

Fig. 1: Light optical images of surfaces after **(a)** boriding **(b)** nitriding **(c)** vanadizing and **(d)** nitrovanadizing

Boriding resulted in formation of a 30-40 µm thick layer, see also SEM image in Fig. 2. At the interface between steel substrate and the boride layer a higher fraction of carbides is observed. This is likely caused by carbon being "pushed" ahead into the alloy by boron. Within the boride layer the presence of primary carbides (characteristic of the Sverker tool steel) is still evident, i.e. by a slight contrast difference in the light optical image. At the interface a zone appearing dark in the light optical image is observed; this is a result of more severe etching of this zone and not e.g. porosity formation. The SEM image reveals some porosity inside the boride layer as discrete voids that appear completely dark in the backscatter image. It is also evident that the boride is rather brittle as cracks have formed near the outer surface; the cracking is likely a result of the embedding and preparation procedure for producing the cross section. The cracking could indicate the presence of FeB, which is known to be more brittle than Fe_2B[2,4]. XRD confirms the presence of Fe_2B in the outermost surface, see Fig. 3. The small peaks observed in the 2 theta range 120-140° may suggest the presence of minor amounts of FeB. The Fe peak indicates presence of some Fe-bcc regions, which could have formed by de-boriding of the surface during the subsequent hardening heat treatment.

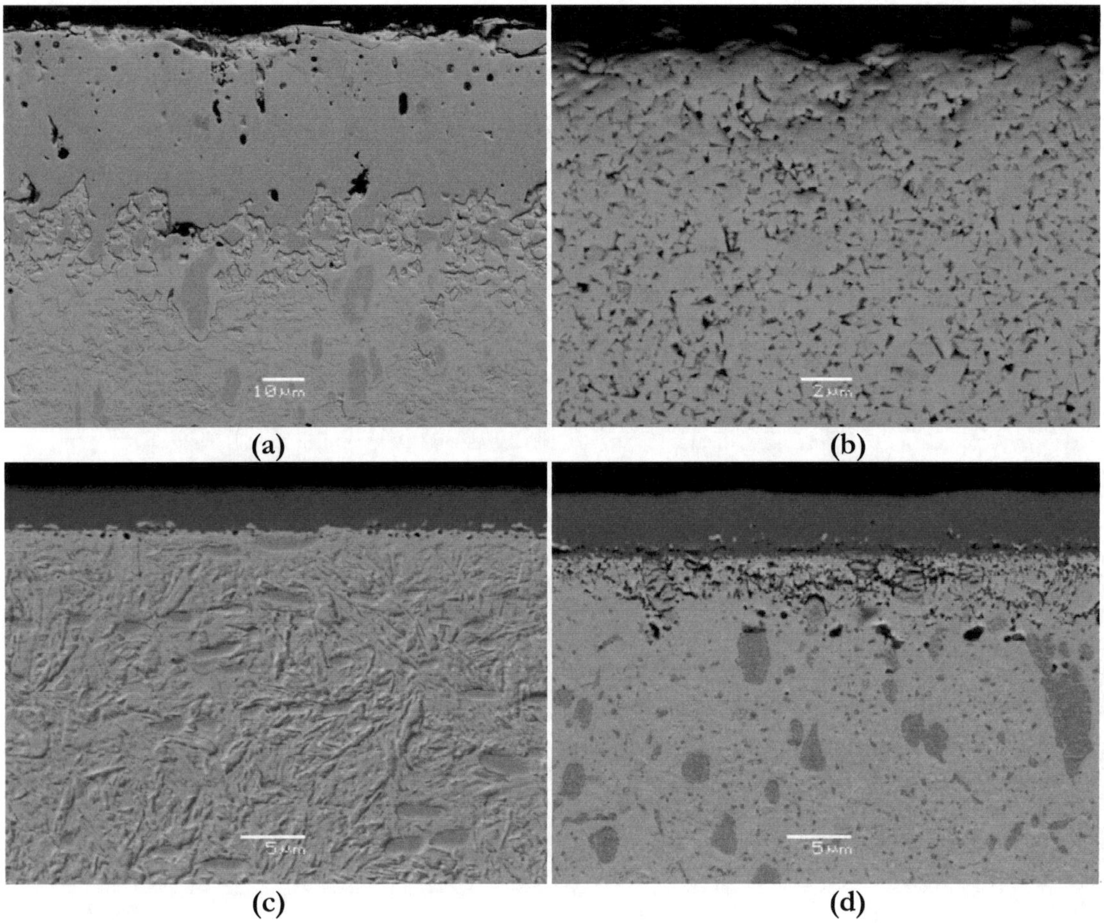

Fig. 2: Backscatter electron images of the **(a)** borided surface and **(b)** cemented carbide reference material and **(c)** vanadized and **(d)** nitrovanadized surfaces

Vanadizing resulted in formation of a thin (2-4 µm) uniform vanadium carbide layer. No major changes in the underlying structure are visible in the light optical or scanning electron images. The X-ray diffractogram confirms the presence of a vanadium carbide (VC) layer, see Fig. 3. Ferrite peaks from the substrate are also clearly observed. Minor peaks from M_7C_3 carbides can be observed too, which can be attributed to the primary carbides present in the substrate.

A compound layer with thickness of approximately 20 µm was formed during nitriding, see Fig. 1. Changes in microstructure are visible below the compound layer, where precipitates have formed at grain boundaries and some carbides have been converted into nitrides; the conversion is visible from the contrast change from light grey to a darker grey (cf. the mechanism described in[11]). Vanadizing after the nitriding procedure (nitrovanadizing) led to formation of a uniform hard layer, which is thicker (4-6 µm) than the VC layer observed on the vanadized sample (2-4 µm). The separate sets of peaks in the diffractogram in Fig. 3 indicate that the layer is a double layer consisting of vanadium carbide (VC) and vanadium nitride (VN). This is consistent with literature where similar layers are found to be carbon rich at the surface and nitrogen rich at the interface[6,8,9].

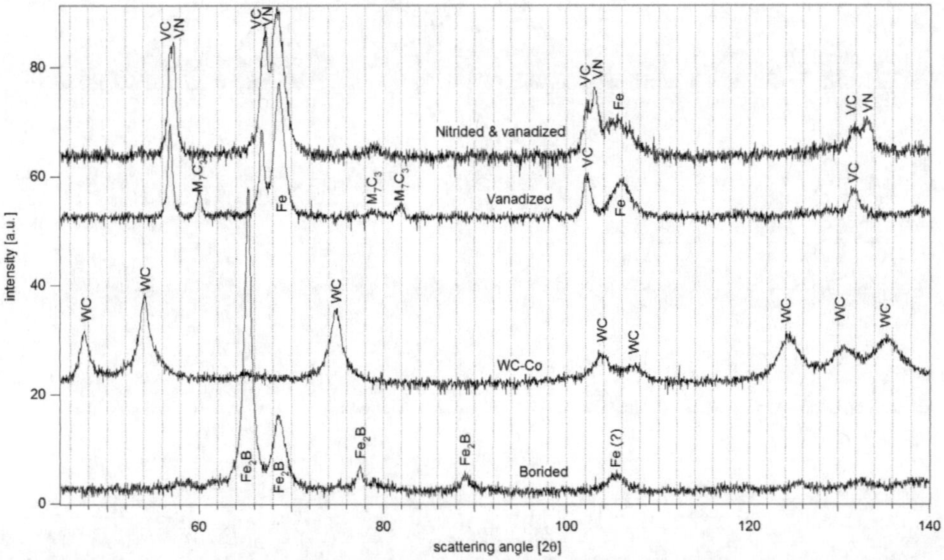

Fig. 3: XRD-diffractograms for WC-Co reference material and borided, vanadized and nitrovanadized samples. Vanadizing has resulted in formation of a VC surface layer. Nitrovanadizing (nitriding followed by vanadizing) has resulted in formation of a double layer consisting of VC and VN. Boriding has resulted in formation of Fe_2B. The small peaks observed in the 2θ range 120-140° may suggest the presence of minor amounts of FeB, while the bcc-Fe peak may indicate a small amount of de-boriding during the final heat treatment step.

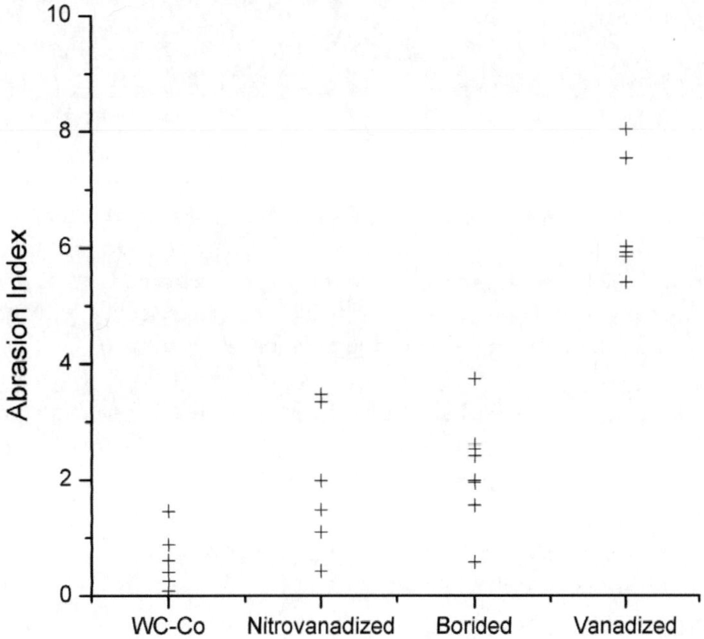

Fig. 4: Results of wear testing. The WC-Co reference material performs the best, while nitrovanadized and borided samples perform clearly better than vanadized and nitrided samples

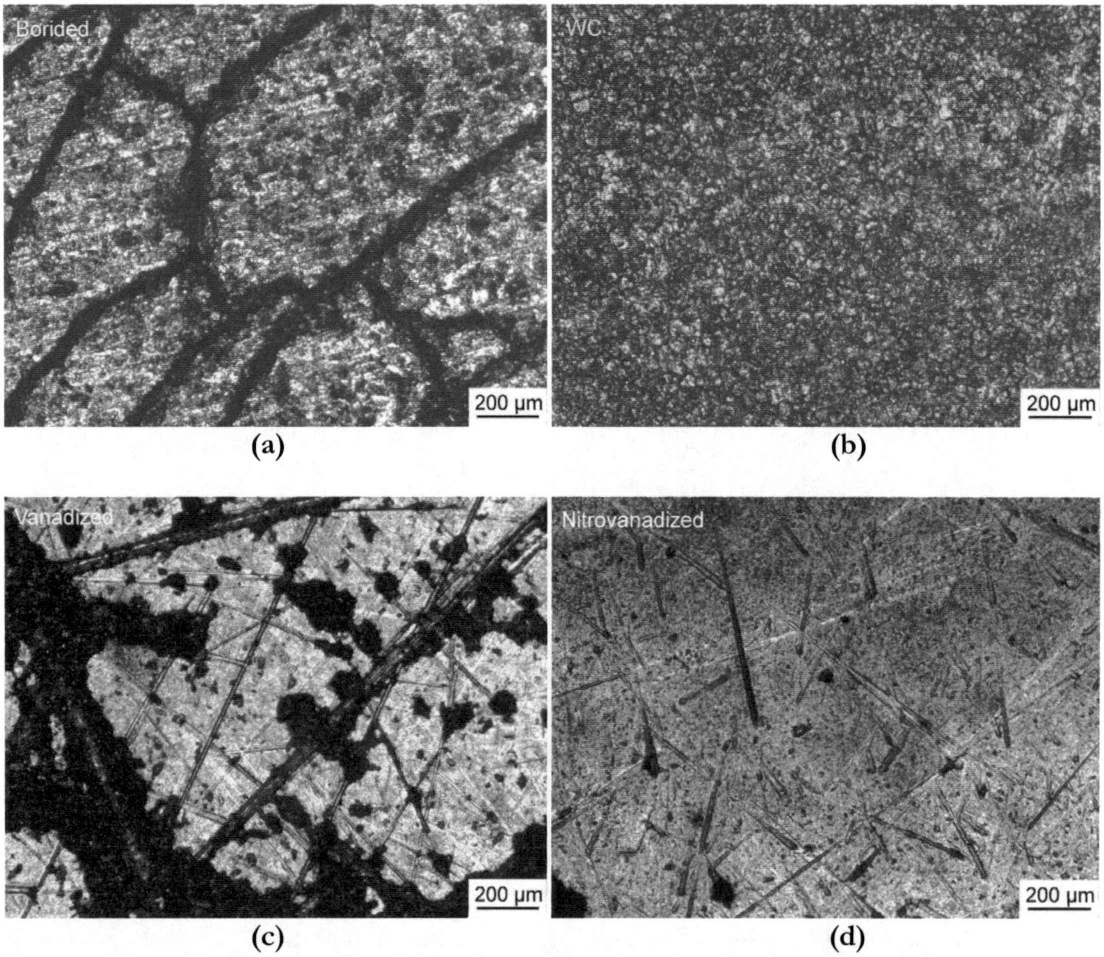

Fig. 5: Light optical images of surfaces after wear testing. For the vanadized samples single quartz particles can perforate the surface layer resulting in delamination and general poor performance. This did not occur for nitrovanadized samples that had a slightly thicker hard surface layer and better support from the substrate (no carbon depleted zone). During initial wear of borided samples, surface cracks were formed but wear rate remained at a level comparable to nitrovanadized samples. For the WC-Co reference material, selective wear of binder phase was observed, but wear rates remained low for the duration of testing.

Close to the bulk alloy porosities are observed within the VC/VN layer and below the layer both porosity formation and precipitation are observed; porosities appear completely dark in the backscatter electron image, Fig. 2. The porosities are likely related to either excess nitrogen in the diffusion zone or incomplete removal of the compound layer, which upon heating can recombine to form N_2 gas. A lower amount of porosities would be expected if nitriding were carried out with a lower nitriding potential, i.e. nitriding conditions that would not lead to formation of a compound layer.

3.2. Wear testing

The results of abrasive wear testing using quartz as an abrasive are shown in Fig. 4. The highest abrasion index and therefore the lowest wear resistance was found for the vanadized samples while the reference cemented tungsten carbide performed the best. Different wear mechanisms were active for the different materials solutions resulting in significant differences in wear performance. For the vanadized samples, the vanadium carbide layer was so thin that the abrasive particles could either directly pierce the VC layer (appearing with bright contrast in the image in Fig. 5) or exert enough force for it to crack and delaminate through buckling. This is clearly visible in Fig. 5 where long perforating wear tracks can be observed on the worn surface of the vanadized sample and at many locations there is no surface carbide layer left. The penetration of the hard surface layer has resulted in a high wear rate due to delamination and due to direct wear of the exposed softer bulk material. In comparison, only few and smaller perforating wear tracks could be observed on the worn surfaces of nitrovanadized samples and the nitride/carbide layer was largely intact after testing. Micro-plowing of quartz particles into the hard surface layer has occurred as visible from the wear tracks in the image in Fig. 5, but it has not lead to the descaling observed for the vanadized samples. This indicates that the wear mechanism is different due to the larger thickness of the hard layer on the nitrovanadized samples. The wear rate of borided samples was comparable to that observed for the nitrovanadized samples but the wear mechanism was different; during initial wear large surface cracks, visible in Fig. 5, formed due to the brittleness of the boride phases and a substantial amount of material has been removed in the vicinity of these cracks. In-between the cracks a more uniform wear of the hard surface occurred without sign of the perforation or plowing tracks observed on the vanadized and nitrovanadized samples. The cemented carbide exhibited the best performance. Since this was not a surfae layer solution and the cemented carbide has good ductility due to the presence of a binder phase, no perforation or cracking was observed resulting in a much more uniform wear as seen from the wear surface shown in figure 5. The softer binder phase has been selectively removed in the near surface but the amount of binder phase in the tested cemented carbide is low (cf. Fig. 2) and no fracture or significant release of hard carbide particles from the surface was observed, thus the wear rate remains low.

In general, it can be said that it is promising that the nitrovanadized samples perform comparably with borided surfaces. This is however only true as long as the layer remains thick enough to avoid penetration and delamination. With further optimisation of the nitriding pre-treatment, to avoid porosity formation and possibly longer vanadizing treatment times to obtain thicker layers, this kind of duplex treatment could hold interest for abrasive wear applications. It must however also be taken into consideration that the borided layers are generally thicker and will therefore last longer with a comparable wear rate. The unfavourable test results observed for the vanadized samples were due to the thin layer formed. It is likely that a thicker vanadium carbide layer would have resulted in a better performance due to a change in wear mechanism from penetration of the layer to a more uniform wear.

4.0. Conclusion

The thermochemical treatments boriding, vanadizing and nitrovanadising resulted in formation of hard surface layers with thicknesses of 30-40μm, 2-4μm and 4-6μm respectively. The nitriding treatment prior to vanadizing influenced the growth kinetics of the V(C,N) layer resulting in a thicker layer compared to vanadizing only. Micro wear testing simulating three-body abrasive wear with quartz as abrasive particles showed that all the treated samples had a higher wear rate than a cemented tungsten carbide reference. The results for nitrovanadizing are promising, since the wear rate is similar to that observed for the borided samples.

5.0. References

1. Sinha, A. K.; ASM Handbook, ASM Intl., 119, vol. 4, p. 437.
2. Sen, U.; Sen, S.; Mater. Charac., 2003, vol. 50, p. 261.
3. Jain, V.; Sundarajan, G.; Surf. Coat. Technol., 2002, vol. 149, p. 21.
4. Sen, S.; Ozbek, I.; Sen, U.; Bindal, C.; Surf. Coat. Technol., 2001, vol. 135, p. 173.
5. Arai, T.; Harper, S.; Handbook, ASM Intl., 1991, vol. 4, p. 448.
6. Chicco, B.; Borbidge, W.E.; Summerville, E.; Mater. Sci. Eng.A, 1999, vol. 266, p. 62.
7. Casteletti, L.C.; Fernandes, F. A. P.; Heck, S. C.; de Oliveira, C. K. N.; Lombardi-Neto, A.; Totten, G. E.; Heat Treating Progress, 2009, vol. 9(5), p. 49.
8. Oliveira, C.K.N.; Benassi, C.L.; Casteletti, L.C.; Surf. Coat. Technol., 2006, vol. 201, p. 1880.
9. Aghaie-Khafri, M.; Fazlalipour, F.; Surf. Coat. Technol., 2008, vol. 202, p. 4107.
10. Jensen, L.R.D.; Fundal, E.; Møller, P.; Jespersen, M.; Wear, 2010, vol. 269, p. 525.
11. Jegou, S.; Barrelier, L.; Kubler, R.; Somers, M.A.J.; HTM - Haerterei-Technische Mitteilungen, 2011, vol. 66(3), p. 135.

Gaseous Surface Hardening of Titanium Alloys

A.E. Barington*, M.S. Jellesen, T.L. Christiansen
Technical University of Denmark.
Produktionstorvet 425
2800 Lyngby.
*Email:alebar@mek.dtu.dk

1.0. Introduction

Titanium has been used for orthopedic implants for decades, due to the material's light weight, good corrosion resistance and biocompatibility. It is also commonly used for other applications such as bone plates and screws for reconstructive surgery. Titanium and its alloys are however known to suffer from poor tribological performance[1]. Traditionally this problem has been solved by simply using other materials for parts intended to be part of tribological systems, such as the popular titanium stem with a CoCr head hip implant design. However, titanium implants are prone to tribocorrosion caused by micro motions, so called fretting wear[2], thus releasing wear particles and ions into the peri implant tissue[3,4]. Surface engineering is one available route to combat this problem[5].

Several distinct approaches for incorporating oxygen into the surface of titanium or a titanium alloy have been employed. Main approaches are anodization, acid treatments, ion implantation, plasma oxidation or thermal oxidation. In a recent review by Spies[6] an overview of oxidation and nitriding via E-beam, laser or plasma for titanium and aluminum alloys is provided. An overview of titanium surface treatments for biomedical applications is also provided by Liu et. al[7], though it does not cover thermal oxidation.

Thermal oxidation of titanium as a detrimental process is a well-researched subject, but less attention has been paid to its use as an intentional surface treatment for titanium[8]. Thermal oxidation is an interesting route for improving the wear and corrosion resistance as well as friction properties due to the ease with which deep layers can be created on parts with complicated geometry[9]. Moreover, there is a large solubility of oxygen in titanium, up to 33.3 at.% and oxygen is a strong (solid solution) strengthener.

Several authors have investigated how oxidation for periods up to an hour at moderate temperatures of around 500 °C improves the corrosion resistance of titanium alloys[10–14]. Other work has found that oxidizing at 700 °C also yields corrosion resistant surfaces[15], however most sources find that oxides created at this temperature are porous and have poor adherence. A mayor issue is that the surface oxide layers tend to delaminate especially if grown at high temperatures or large thicknesses. This is especially problematic from a fatigue point of view and studies which investigate this characteristic often find that thermally oxidized samples have significantly worse fatigue properties than untreated samples[16–18]. It was however shown that treatments which produced a surface with minimal oxide formation and instead oxygen in solid solution the detrimental effect of oxidation on fatigue resistance is minimal[17,19], while the improvements to wear and corrosion resistance are maintained.

One method that has been investigated intensely for this purpose is oxygen boost diffusion, where a 2 stage procedure is employed. First a short oxidation, typically in air, creates an oxide layer a couple μm thick, and then a subsequent vacuum annealing dissolves this layer into solid solution oxygen and various suboxides. This process has been evaluated for both Ti6Al4V[20–22], pure Ti[22–25] and other alloys[26]. It is noted that the extended annealing results in rampant grain coarsening for the pure titanium samples, e.g. grain sizes exceeding 100μm[22].

Alternatively, oxidizing can be performed in a low pressure system to limit the oxygen activity[27,28]. By oxidizing in an unspecified media, at pressures between 6.6 and 66 mPa, for durations of 1 to 5 hours and temperatures of 650 to 750 °C, thick oxygen rich diffusion layers can be achieved.

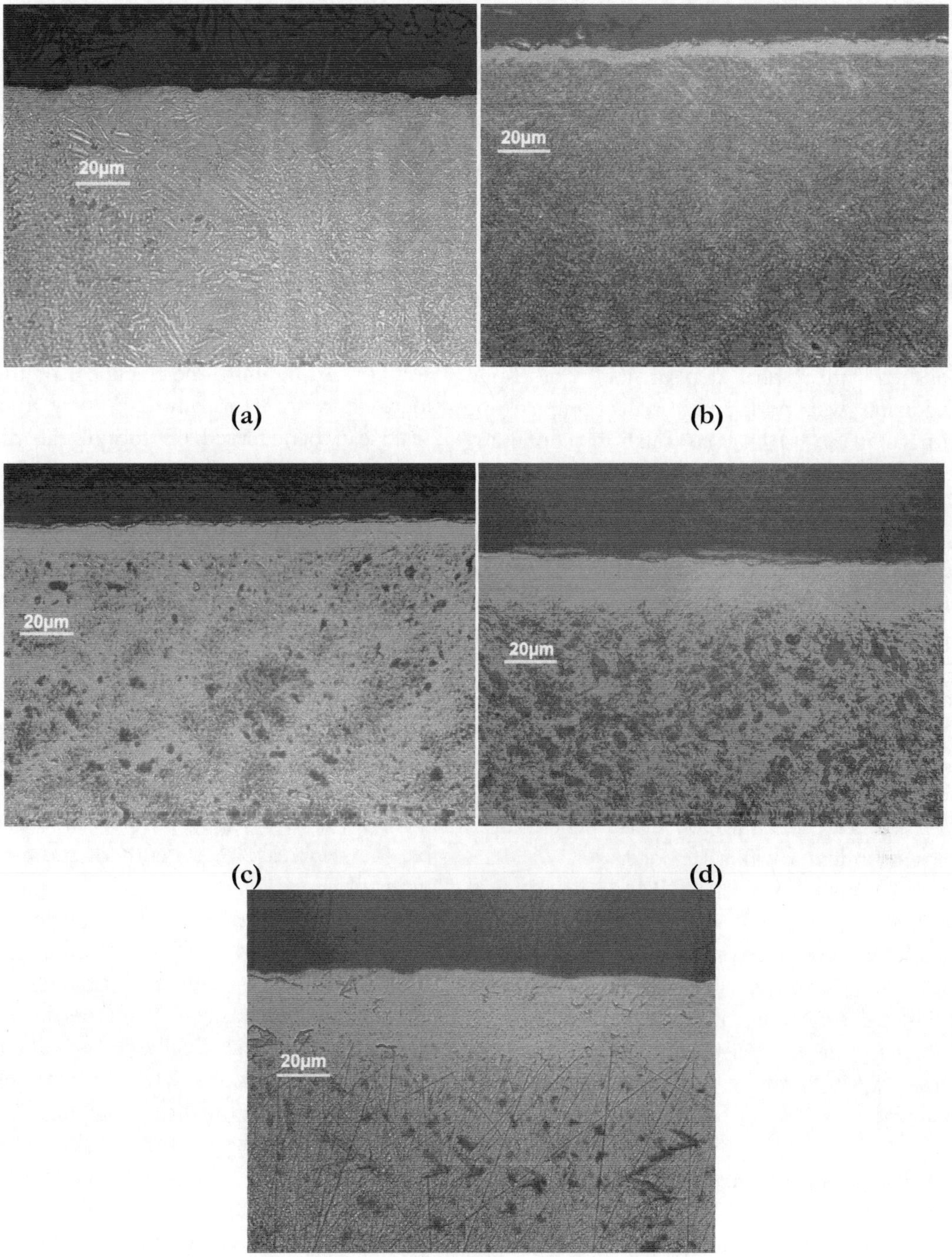

Fig. 1: Cross sections of treated samples using light optical microscopy. Images a through e represent increasing temperature, such that **(a)** 550ºC **(b)** 600ºC **(c)** 650ºC **(d)** 700ºC and **(e)** 750 ºC.

In this paper the possibility of creating layers of oxygen in solid solution in titanium using dilute oxygen gas sources at atmospheric pressure is investigated in abroad temperature interval. Using this approach rather than those previously described has the advantage of not requiring a tight vacuum seal or a multistage process.

2.0. Experimental

Samples of grade 1titanium were polished with a final step of 4000 grit paper. Oxidizing was performed in a ceramic pipe furnace in an O_2-N_2 gas mixture, containing 0.21 vol.% O_2.The gas flow rate was held constant during all treatments at 0.5 l min^{-1} and a total treatment time of 17 hours. The temperature was varied from 550 to 750 °C in increments of 50 °C.

Cross sections of the samples were embedded in an epoxy resin and ground with a final polishing step of colloidal silica. In order to reveal the microstructure of the cross sections Kellers reagent was applied (20 seconds).

Microhardness profiles were obtained with a Futuretech FM-700 instrument using a Vicker's indenter with a 10g load.

X-ray diffraction (XRD) analysis was performed using a Bruker AXS D8 Discover instrument equipped with a Cu kα source in locked couple mode. This setup leads to 90% of the signal originating from the top 3.7 μm of material. For the samples with loose oxides these were carefully maintained for XRD analysis.

3.0. Results and interpretation

Light optical micrographs of the samples are shown in Fig. 1 and the thickness of the surface hardened zone, average of three measurements is shown in Fig. 2.

A fragile surface layer could easily be scraped off as a fine white powder from the samples treated at 700 °C and 750 °C. Below this layer the samples have irregular interference colorings ranging from yellow to blue. This loosely adherent layer was destroyed in preparing the samples for cross sections and is not included in the layer thickness' shown on Figs. 1 and 2. All the other samples achieved a dark gray color after treatment that appeared independent of temperature. The 550 °C sample had an orange sheen, while the other samples appeared matte.

The microstructure (cf. Fig. 1) consists of a dominant diffusion zone of oxygen in solid solution in titanium; the thickness of this zone increases with temperature. A very thin oxide scale can beobserved on top of the diffusion zone.

Microhardness depth profiles are presented in Fig. 3. A substantial hardness increase is obtained from treatments carried out at 600 °C and above. No noticeable change in hardness was observed in the 550 °C sample at depths of 5 μm or deeper. The 750 °C sample exhibits a hardness plateau to a depth of around 15 μm, likely indicating oxygen saturation to this depth. The hardness depth profiles reflect the diffusion profiles of oxygen, i.e. oxygen in solid solution. At the highest temperature of 750°C a hardness increase to a depth of more than 40 μm is obtained.

X-ray diffractograms are given in Fig. 4. The treated samples all displayed rutile TiO_2 peaks, as well as expanded α-titanium. Here it should be emphasized that the information depth is relatively shallow and this will make the surface adjacent zone (viz. oxide scale)appear more dominant in the diffractograms.

The samples treated at 650°C and above show prominent peaks of α-Ti (alternatively Ti_6O which is an ordered solid solution of oxygen in Ti). In addition peaks located at about 2θ=51, 61, and 73° are also visible at samples treated at 650 °C and above. These peaks are most likely ε-Ti_2N. At 70° reflections representing rutile and α-titanium overlap making unambiguous identification of these phases difficult.

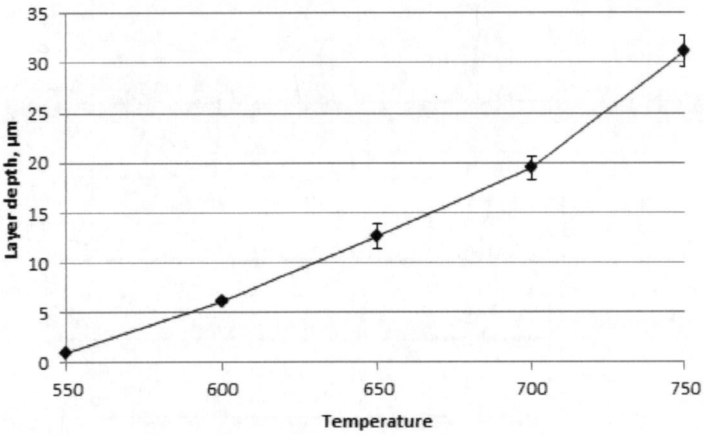

Fig. 2: Layer depth of adherent surface layers, error bars indicate one standard deviation

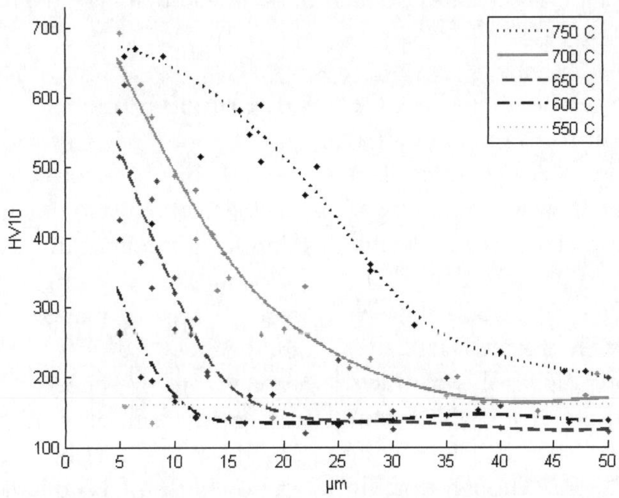

Fig. 3: Microhardness profiles of the treated samples

4.0. Discussion

By oxidizing in a low oxygen environment at atmospheric pressure hard layers of oxygen in solid solution can be created. However, under the applied conditions it is not possible to fully prevent oxide formation. It has been shown that between 600 and 650 °C layers of mostly oxygen in solid solution that are tens of µm deep are formed. Also using the same methods even deeper and harder layers can be created using higher temperatures, but with the development of a brittle oxide scale. This scale is loosely adherent and easily removable.

The layer depths shown in Fig. 2 correspond to a measured microhardness of above 250 to 300 HV10g regardless of temperature. This hardness is higher than the bulk hardness of 160 HV10g, thus the depths measured in Fig. 2 are not the actual penetration depths but a threshold of oxygen content above which additional corrosion protection is acquired.

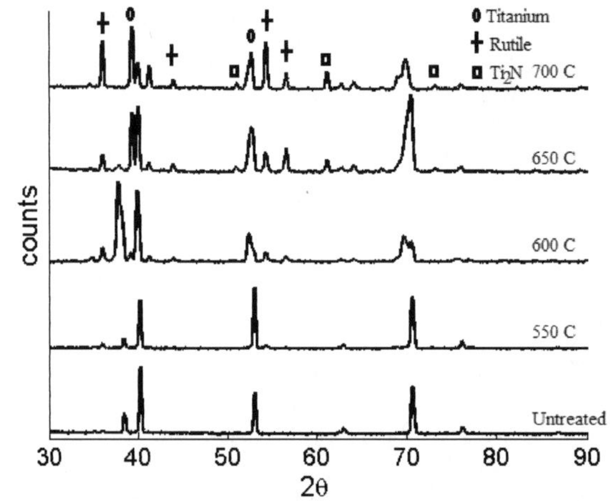

Fig. 4: X-ray diffractograms. Reflections uniquely corresponding to a single phase are indicated in the figure. The Ti6O phase has expected peaks at the same locations as those marked as titanium.

The 0.21% oxygen content of the treatment gas is equivalent to an oxygen partial pressure of around 210 Pa. Hertl et. al[23]applied dilute oxygen environment, using even lower oxygen partial pressures. Here it was found that at 730 °C loosely adhering oxides were formed already at 30 Pa, and at 660 °C well adhering oxide layers were formed at 1 Pa, though not at 0.2 Pa and below. This conflicts with the findings of Fedirko et.al[27] who found that deep layers could be developed at pressures as low as 0.007 Pa. Though notably longer times were employed in the present study similar results are obtained, likely due to the formation of a rutile layer serving as a diffusion barrier to prevent uptake of additional oxygen, early in the treatment process.

The experiments were conducted using nitrogen as a diluting gas, which resulted in the formation of titanium nitrides. This means that the obtained solid solution most likely include both oxygen and nitrogen atoms.

5.0. Conclusions

It has been shown that it is possible to developthick layers containing oxygen in solid solution in titanium in a single step process at atmospheric pressure. Oxygen in solid solution was obtained in the temperature range of 600 to 750 °C. XRD patterns also revealed the presence of rutile TiO_2as a thin surface adjacent layer. At temperatures of 650 °C and above Ti_2N was also formed by reaction with the N_2 diluting gas.

The treatments resulted in hardness values up to 700 HV10g in the oxygen diffusion zone, with gradual hardness decrease to a depth of more than 40 μm, with most of this layer composed of oxygen in solid solution. This depth and hardness might be sufficient to protect against fretting corrosion failure against materials commonly mated against titanium in orthopedic implants such as bone, bone cement and CoCr alloys.

6.0. References

1. Khan, M. A.; Williams, R. L.; Williams, D.F.; Biomaterials, 1996, vol. 17, p. 2117.
2. Waterhouse, R. B.; Wear, 1984, vol. 100, p. 107.
3. Hoeppner, D.; Chandrasekaran, V.; Wear, 1994, vol. 73, p. 189.
4. Fayeulle, S.; Blanchard, P.; Vincent, L.; Tribol. Trans., 1993, vol. 36, p. 267.
5. Fu, Y.; Wei, J.; Batchelor, A. W.; J. Mater. Process. Technol., 2000, vol. 99, p. 231.
6. Spies, H.-J.; Surf. Eng., 2010, vol. 26, p. 126.
7. Liu, X.; Chu, P.; Ding, C.; Mater. Sci. Eng. 2004, vol. R47, p. 49.
8. Zhecheva, A.; Sha, W.; Malinov, S.; Long, A.; Surf. Coatings Technol., 2005, vol. 200, p. 2192.
9. Browne, M.; Gregson, P.; Biomaterials, 1994, p. 894.
10. Ashrafizadeh, A.; Ashrafizadeh, F.; J. Alloys Compd., 2009, vol. 480, p. 849.
11. Chrzanowski, W.; J. Achiev. Mater. Manuf. Eng., 2006, vol. 18, p. 67.
12. Kumar, S.; Sankara Narayanan, T. S. N.; Ganesh Sundara Raman, S.; Seshadri, S. K.; Mater. Chem. Phys., 2010, vol. 119, p. 337.
13. Jamesh, M.; Sankara Narayanan, T. S. N.; Chu, P. K.; Mater. Chem. Phys., 2013, vol. 138, p. 565.
14. Biswas, A.;Dutta Majumdar; J. Mater. Charact., 2009, vol. 60, p. 513.
15. Shankar, A. R.; Karthiselva, N. S.; Mudali, U. K.; Surf. Coatings Technol., 2013, vol. 235, p. 45.
16. Leinenbach, C.; Eifler, D.; Biomaterials, 2006, vol. 27, p. 1200.
17. Ebrahimi, A. R.; Zarei, F.; Khosroshahi, R. A.; Surf. Coatings Technol., 2008, vol. 203, p. 199.
18. Leinenbach, C.; Eifler, D.; Acta Biomater., 2009, vol. 5, p. 2810.
19. Streicher, R. M.; Weber, H.; Schön, R.; Semlitsch, M.; Biomaterials, 1991, vol. 12, p. 125.
20. Dong, H.; Li, X. Y.; Mater. Sci. Eng. A, 2000, vol. 280, p. 303.
21. Zhang, Z. X.; Dong, H.; Bell, T.; Xu, B.; J. Alloys Compd., 2007, vol. 431, p. 93.
22. Zabler, S.; Mater. Charact., 2011, vol. 62, p. 1205.
23. Hertl, C.; Werner, E.; Thull, R.; Gbureck, U.; Biomed. Mater., 2010, vol. 5, p. 054104.
24. Arslan, E.; Totik, Y.; Demirci, E.; Alsaran, A.; J. Mater. Eng. Perform., 2010, vol. 19, p. 428.
25. Takamura, A.; Trans. Japan Inst. Met., 1962, vol. 3, p. 10.
26. Boettcher, C.; Surf. Eng., 2000, vol. 16, p. 148.
27. Fedirko, V. M.; Pichuhin, A. T.; Luk'yanenko, O. H.; Onuferko, V.S.; Mater. Sci., 2005, vol. 41, p. 208.
28. Fedirko, V. M.; Pohrelyuk, I. M.; Yas'Kiv, O. I.; Mater. Sci., 2006, vol. 42, p. 299.

Keyword Index

Author Index